Gamma Lines

Asian Mathematics Series
A Series edited by Chung-Chun Yang
Department of Mathematics, The Hong Kong University of Science and Technology, Hong Kong

Gamma Lines
On the Geometry of Real and Complex Functions

Grigor A. Barsegian

*Institute of Mathematics, National
Academy of Sciences of Armenia
Yerevan, Republic of Armenia*

Routledge
Taylor & Francis Group

LONDON AND NEW YORK

First published 2002 by Taylor & Francis

2 Park Square, Milton Park, Abingdon, Oxfordshire OX14 4RN
52 Vanderbilt Avenue, New York, NY 10017

Routledge is an imprint of the Taylor & Francis Group, an informa business

First issued in paperback 2019

British Library Cataloguing in Publication Data
A catalogue record for this book is available
from the British Library

Library of Congress Cataloging in Publication Data
A catalog record for this book has been requested

ISBN 978-0-367-39591-9

Contents

Introduction to the series

The *Asian Mathematics Series* provides a forum to promote and reflect timely mathematical research and development from the Asian region, and to provide suitable and pertinent reference or text books for researchers, academics and graduate students in Asian universities and research institutes, as well as in the West. With the growing strength of Asian economic, scientific and technological development, there is a need more than ever before for teaching and research materials written by leading Asian researchers, or those who have worked in or visited the Asian region, particularly tailored to meet the growing demands of students and researchers in that region. Many leading mathematicians in Asia were themselves trained in the West, and their experience with Western methods will make these books suitable not only for an Asian audience but also for the international mathematics community.

The *Asian Mathematics Series* is founded with the aim to present significant contributions from mathematicians, written with an Asian audience in mind, to the mathematics community. The series will cover all mathematical fields and their applications, with volumes contributed to by international experts who have taught or performed research in Asia. The material will be at graduate level or above. The book series will consist mainly of monographs and lecture notes but conference proceedings of meetings and workshops held in the Asian region will also be considered.

Preface

The history of mathematics is in a considerable extent connected with the study of solutions of the equation $f(x) = a = const$ for functions $f(x)$ for one real or complex variable. However we knew surprizingly little about solutions of $u(x, y) = t = const$ for functions of two real variables. These solutions, called level sets, are very important because of their applications in physics, environmental problems, biology, economics, etc., as they mean a "map" of an appropriate process described by the function $u(x, y)$ for given parameters (x, y).

This book studies the concept of Γ-lines that generalizes both the classical concepts of level sets and a-points. The author aims to show how large is the field of possible applications of Γ-lines and to present a book accessible for specialists in different sciences; at least to read Chapters 1, 3, 5 one need to be familiar only with beginning of the standard course of complex analyses.

An interesting circumstance is that by reading these formally quite simple chapters one can particularly get idea about leading ideas and main conclusions of classical Value distribution theory of R. Nevanlinna that study mainly numbers of a-points of meromorphic functions in the complex plane. Moreover, constructed theory of *Distribution of Γ-lines* permit to transfer Value distribution theory to a new stage of problems in studying the geometry of a-points instead of the numbers of a-points, considered in the classical theory.

Meromorphic functions in the unit disk is less investigated. For functions of slow growth the Second fundamental theorem of Value distribution theory do not give efficient consequences. This is an essential gap in complex function theory since many known classes of functions (particularly classes of Nevanlinna, Dirichlet, Hardy) do have slow growth. The *Γ-lines approach* offered in this book permits to describe the numbers as well as the locations of a-points of all functions meromorphic in the unit disk.

On the other hand, we study solutions of the following system $u(x, y) = t = const$, $|grad\ u(x, y)| = R = const$, for some classes of functions $u(x, y)$ associated with meromorphic functions. The existence of the solutions for different classes of $u(x, y)$ is studied in the theory of Free Boundary Problems which thanks to rich applied content is one of the recent hot topics. Using Γ-lines we are able to describe the numbers of these solutions for a class of functions $u(x, y)$. To our surprise we find a kind of Nevanlinna Deficiency Relation for this distinct topic.

The Γ-lines of some particular functions classes as well as applications of level sets in physics and engineering were subjects of very active international research in the last 15 years. The author hopes that the methods of the present book will be useful for the development of the mentioned and of new topics related both to pure mathematics and applications. At first this concern a new and wide program: the study of level sets for solutions of partial differential equations (and other type equations). Physical interpretations of these level sets and respectively expected bearings in applications, seem to be evident.

My pleasant duty is thanking people for their attention and valuable discussions while attaining the book results: Professors C. Andreian-Cazacu, N. Arakelian, A. Beliy, W.H.J. Fuchs, A. Goldberg, A. Gontchar, S. Mergelian, G. Suvorov.

I am greatly indebted to Professors H. Begehr and I. Laine for carefully reading the book manuscript and suggesting several improvements.

My special thanks go to Professor C.C. Yang and Dr G. Sukiasian for their patience and invaluable help during the final stage of preparation of this book.

YEREVAN, 1999

INTRODUCTION

The theory of Γ-lines deals with numerous classical problems of contemporary complex analysis, physics and engineering. Since some particular cases of the Γ-lines become level sets, the theory of Γ-lines leads to a "theory of level sets". Even though the notion of level sets is widely discussed in many applied sciences, no theory has been constructed.

Particularly, theory of Γ-lines generalizes the main conclusions of the classical theories of Nevanlinna and Ahlfors. It extends the concepts and the main conclusions of these theories and also offers some new ways of their further development. Some of the results of this theory are new principles, since they are valid for very large classes of functions.

For physics and engineering, the theory of Γ-lines can offer new approaches in the study of level sets of physical magnitudes which one can observe everywhere in Nature and in theoretical investigations, especially in the theory of Catastrophes.

The author aims to present a book on Γ-lines accessible for experts in all mentioned fields. The short survey below shows the mathematical and physical aspects of the theory of Γ-lines.

An essential part of the complex analysis, particularly the classical theories of Nevanlinna and Ahlfors, deals with the number of zeros or, more generally, the number of a-points of the functions $w(z)$, analytic or meromorphic, in given complex domains D or in the complex plane C. The a-points of $w(z)$ are the solutions of the equation $w(z) - a = 0$ or, which is the same as the pre-images $w^{-1}(a)$ of the mapping given by $w(z)$.

Let us consider the solutions

$$u(x,y) - t = 0, \quad u(x,y) = \operatorname{Re} w(z),$$

where t is a real constant. These solutions actually are the pre-images $w^{-1}(\Gamma)$, where Γ is a straight line $\{w : \operatorname{Re} w = t\}$. Let us now consider the more general case $w^{-1}(\Gamma)$, where Γ is an arbitrary smooth Jordan curve in the complex plane. We call Γ-lines the pre-images $w^{-1}(\Gamma)$ for the mapping $w(z)$ in a given domain D. Obviously, this concept to some extent generalizes the classical concept of a-points, i.e. $w^{-1}(a)$. Thus, we have a reason to develop a theory of Γ-lines similar to the classical theories describing a-points. However, there are numerous other reasons why construction of such a theory can be useful and important. First of all, we note that any result describing Γ-lines of an analytic function w describes, in fact, important applications in harmonic functions $\operatorname{Re} w$ if $\Gamma = \{w := \operatorname{Re} w = t\}$. Therefore, we deal with level sets of real functions $\operatorname{Re} w$ or $|\operatorname{grad} u|$ if $\Gamma = \{w : |w| = t\}$ or $\Gamma = \{w' : |w'| = t\}$. In mathematics Γ-lines offer some new approaches in the investigation of some old and new problems: among them, problems related to Gelfond quantities, to asymptotic behaviors, to an old idea of Nevanlinna concerning the description of Riemann surfaces. Bearing of Γ-lines in the study of locations of a-points deserves to be mentioned especially since it fills an essential gap in the theory of complex functions. The preceding classical results concerning the a-points, particularly Nevanlinna and Ahlfors' theories, mainly describe the numbers of the a-points of functions w, where w are either arbitrary functions meromorphic in the complex plane or functions meromorphic in the unit disk, having "fast" growth. The main conclusions of these theories are not more true for functions in the unit disk with "slow" growth.

Obviously, the study of the locations of the a-points is the next stage of development of the theory of meromorphic functions. The theory of Γ-lines offers a tool for investigating the locations of the a-points. This theory permits also the consideration of meromorphic functions of "slow" growth in the unit disk, and even some classes of non-analytic functions. Moreover it allows to describe the value distribution and locations of a-points for the known classes of functions as H^p, Dirichlet class, bounded functions, and Blaschke products in terms of characteristics of these classes.

As to functions meromorphic in the complex plane, the book presents a rather complete theory of "distribution of Γ-lines" similar to Nevanlinna Value Distribution Theory.

The study of Γ-lines seems to be promising since there are different branches of physics where these lines can be interpreted as *level sets of physical processes*. Indeed, if a function $\operatorname{Re} w(z)$ describes a physical phenomenon in a planar point z (for instance, its velocity, electrical or magnetic tension, temperature, level of radiation, etc.), then for $\Gamma = \{w : \operatorname{Re} w = t\}$ the Γ-lines mean those sets on the plane where these magnitudes are equal to a given constant t. For example, if $\operatorname{Re} w(z)$ describes the temperature of the sea surface (the planar case) and $\Gamma = \{w : \operatorname{Re} w = 0\}$, then Γ-lines are the boundaries of icebergs. Another example: Let $\operatorname{Re} w(z)$ describe the temperature in a nuclear process and T be the critical temperature of materials' transition to plasmatic state. Then Γ-lines mean the boundary of the plasmatic state if $\Gamma = \{w : \operatorname{Re} w = T\}$. Similarly, Γ-lines can mean the boundaries of domains of dangerous radiation in the investigations of environmental problems or the boundaries with critical velocities, tensions, etc. in physics and engineering of catastrophes.

Thus Γ-lines have important interpretations in physics unlike a-points which are generalized by Γ-lines. Therefore, any result described by these lines can be interpreted as a result in a corresponding field of physics. The study of Γ-lines also opens another prospect for applications of Nevanlinna theory in physics. This is due to some explicit connections between Nevanlinna theory and the theory of Γ-lines presented in this book.

Among the objects connected to level sets we consider also solutions of the following system $u(x,y) = t = $ const, $|\text{grad } u(x,y)| = R = $ const, for some classes of functions $u(x,y)$ associated with meromorphic functions. The existence of the solutions for different classes of $u(x,y)$ is studied in the theory of Free Boundary Problems which, thanks to the rich applied content, is one of the recent hot topics. It is easy to understand that while the first equation in this system shows, so to say, allocation of a physical process, the second equation shows its propagation. Using Γ-lines we give a description for the numbers of these solutions that surprisingly resemble the known Nevanlinna Deficiency Relations.

The main results of the theory of Γ-lines for functions meromorphic in the complex plane or in the unit disk and even for some more general classes of non-analytic functions were established by the author in 1977–1978. Since then the author believes that the theory of Γ-lines has good prospects. Clearly, the author is pleased that many other specialists have joined in the investigations of these lines. The Γ-lines of the specific class of univalent functions have first been considered by Hayman and Wu in 1981 and later by Garnett, Jones, Bishop, Carleson, Gehring, Fernandez, Heinonen, Martio.

Also the author hopes that the methods developed in the book permit transfer to another new and wide program: to the study of level sets for solutions of partial differential equations (and other type of equations). Physical interpretations of these level sets and correspondingly the expected bearings in applications seem to be evident.

CHAPTER I

TANGENT VARIATION PRINCIPLE: SATELLITE PRINCIPLES

1.1. Modifications of Length–Area Principle: Connection with Various Classes of Functions

1.1.1. The Ahlfors' classical length–area principle (Ahlfors [1], 1930). It is one of the remarkable relations of the theory of functions, unique in its generality and clearness. According to this principle, for any function $w(z)$ regular in a domain D the following inequality is true:

$$\int_0^\infty \frac{L^2(D.\Gamma(R))}{Rp(R)} dR \le 2\pi S(D), \tag{1.1.1}$$

where $L(D,\Gamma(R))$ is the sum of the lengths of the curves in D on which $|w(z)| = R$ (which is the same as the curves $w^{-1}(\Gamma(R)))$, where $\Gamma(R)$ is the circle $\{w : |w| = R\}$ and

$$p(R) = \frac{1}{2\pi} \int_0^{2\pi} n\left(D, Re^{i\vartheta}\right) d\vartheta,$$

where $n\left(D, Re^{i\vartheta}\right)$ is the number of roots of the equation $w(z) = Re^{i\vartheta}$ in D, according to their multiplicities, and $S(D)$ is the area of D.

This important connection between these different magnitudes has numerous effective applications in the theory of univalent functions, quasiconformal mappings and other topics (see Lelong-Ferrand [1] and Suvorov [1]).

The inequality (1.1.1) was the initial point for the study of p-valent functions in the unit disk, the circumstantial or areal mean of p-valent functions. An inconvenience arising from the fact that $p(R)$ appears under the integral (1.1.1) in general case is eliminated for such functions. For instance, if a function is p-valent in mean on the unit circle, then $p(R) \le p = $ const, and hence

$$\int_0^\infty L^2\left(D.\Gamma(R)\right)\frac{dR}{R} \le 2\pi p S(D). \tag{1.1.1'}$$

A series of remarkable investigations by Cartwright, Spencer and Hayman based on the inequalities (1.1.1) and (1.1.1') are summarized in the monograph by Hayman [1].

1.1.2. Some modifications of the length–area principle.

Assuming $z = re^{i\varphi}$ denote

$$S_n(D) = \int\int_D |(w'(re^{i\varphi})|^n \, rdr d\varphi, \quad n = 1, 2,$$

$$A_n(D) = \int\int_D \rho^n(re^{i\varphi})rdr d\varphi, \quad n = 1, 2, \quad \rho(re^{i\varphi}) = \frac{|w'(re^{i\varphi})|}{1 + |w(re^{i\varphi})|^2}.$$

Here $S_2(D)$ is the area of the w-image of D, counting the multiplicity of covering, and $A_2(D)$ is the spherical area of the same w-image, again counting the multiplicity of covering.

Below, we give two new inequalities which can be considered as modifications of the length–area principle due to their geometric meaning. The following relation is true for any function w regular in a domain D:

$$\int_0^\infty L(D, \Gamma(R))dR = S_1(D) \leq (S_2(D) \cdot S(D))^{1/2}. \qquad (1.1.2)$$

Besides, if w is a function meromorphic in D, then the following inequality holds:

$$\int_0^\infty \frac{L(D, \Gamma(R))}{1 + R^2}dR = A_1(D) \leq (A_2(D) \cdot S(D))^{1/2}. \qquad (1.1.3)$$

The inequalities (1.1.2) and (1.1.3) are more simple than (1.1.1). They establish connections between some estimates of integrals of L and the mentioned classical integrals. Some well-known classes of functions, including that of Dirichlet ($S_2(D) < \infty$) and that of bounded spherical area ($A_2(D) < \infty$), are defined in terms of these classical integrals.

We suppose that inequalities (1.1.2) and (1.1.3) can be used for the study of the above mentioned classes of functions in the same way as (1.1.1) was used for p-valent functions. For starting and solving such problems, one can follow Hayman's scheme [1] which applies inequality (1.1.1) for the study of p-valent functions. For instance, it is interesting to find out to what extent the results of Hayman [1] are true for the mentioned classes.

The following relations give several connections between some integrals of L and other well-known magnitudes. Let w be a function regular in the unit disk. If D is the disk $D(r) = \{z : |z| < r\}$, we shall write $L(r, \Gamma)$ instead of $L(D(r), \Gamma)$. Denote

$$J_p(r, w) = \frac{1}{2\pi} \int_0^{2\pi} |w(re^{i\varphi})|^p \, d\vartheta, \quad p > 0,$$

$$L(r, w) = \int_{|z|=r} \rho(re^{i\varphi})rd\varphi.$$

For the case $p > 1$, (1.1.2) and Hölder's inequality imply

$$\int_0^\infty L(r, \Gamma(R))dR = S_1(r) \le 2\pi \int_0^r J_p^{1/p}(t, w')tdt, \qquad (1.1.2')$$

$$\int_0^\infty L(r, \Gamma(R))dR = S_1(r) \le \left(\int\int_{D(R)} |w'|^p d\sigma\right)^{1/p} (S(r))^{1-1/p}. \qquad (1.1.2'')$$

From (1.1.3) we derive

$$\int_0^\infty \frac{L(r, \Gamma(R))}{1+R^2}dR = A_1(r) = \int_0^r L(t, w)dt \le (A_2(r)S(r))^{1/2}. \qquad (1.1.4)$$

The last relations are of interest as the function w is assumed there to be of the H_p class if $\sup_r J_p(r, w) < \infty$. and of the Tsuji class if $\sup_r L(r, w) < \infty$. The inequality (1.1.4) is interesting also as the magnitude $A_2(r)/\pi$ coincides with the classical Ahlfors' characteristic $A(r)$.

The following relation connects some double integrals of L and the magnitudes J_p for any given $p > 0$:

$$\frac{p^2}{2\pi^2} \int_0^r \left(\int_0^\infty \frac{L(t, \Gamma(R))}{R^{1-\frac{p}{2}}}dR\right)^2 \frac{dt}{t^3} \le J_p(r, w) - J_p(0, w). \qquad (1.1.5)$$

For proving inequalities (1.1.2)–(1.1.5) we need the following.

1.1.3. Basic identity. Let $w(z)$ be a function meromorphic in a domain D, and let $\Psi(R)$ be a continuous function for $0 \le R < \infty$, such that $\Psi(R) > 0$ as $R > 0$. Then

$$\int_0^\infty \frac{L(D, \Gamma(R))}{\Psi(R)}dR = \int\int_D \frac{|w'(z)|}{\Psi(|w(z)|)}d\sigma, \qquad (1.1.6)$$

where $d\sigma$ is the area element.

For further applications it is necessary first to consider the case where Γ is an arbitrarily smooth Jordan curve.

Proof. In the following constructions we shall exclude from D all zeros, poles and multiple points of the function $w(z)$. We represent the set of lines

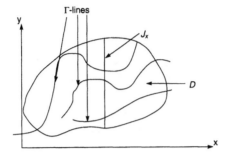

Figure 1.1

$w^{-1}(\Gamma) \cap D$ as the sum $l_x(\Gamma) \cup l_y(\Gamma)$, where $l_x(\Gamma)$ is the totality of the arcs from $w^{-1}(\Gamma) \cap D$, such that in any point belonging to $l_x(\Gamma)$ the smaller angle α_x between the tangent and axis x is less than or equal to $\pi/4$; $l_y(\Gamma)$ is the totality of the arcs from $w^{-1}(\Gamma) \cap D$, such that in any point belonging to $l_y(\Gamma)$ the smaller angle α_y between the tangent and axis y is less than $\pi/4$ (see Figures 1.1 and 1.2).

Let $L_x(D,\Gamma)$ and $L_y(D,\Gamma)$ be the total lengths of the arcs $l_x(\Gamma(R))$ and $l_y(\Gamma(R))$, respectively. Besides, let D_x (or (D_y)) be the set of those $z \in D$ for which the arc $l(\Gamma(|w(z)|))$ through z is of the type $l_x(\Gamma(|w(z)|))$ (or $l_y(\Gamma(|w(z)|))$). It is obvious that

$$L_x(D,\Gamma) + L_y(D,\Gamma) = L(D,\Gamma) \tag{1.1.7}$$

and

$$D_x \cup D_y = D. \tag{1.1.8}$$

For a fixed x, by J_x we denote the set $D \cap \{z : \operatorname{Re} z = x\}$, by $\Phi(D,x,\Gamma)$ the number of points $z_{i,x}(\Gamma)$ of the intersection of J_x with the sum of arcs $l_x(\Gamma)$, and by $\alpha_{i,x}(\Gamma)$ the smaller angle between the axis x and the tangent to the arc from $l_x(\Gamma)$ at the point $z_{i,x}(\Gamma)$. Note that $\alpha_{i,x}(\Gamma) \le \pi/4$ due to the definition of arcs $l_x(\Gamma)$. Hence for the length element Δl of an arc from $l_x(\Gamma)$ the following relation is true: $\Delta l \sim \Delta x / \cos \alpha_{i,x}(\Gamma)$ at the point $z_{i,x}(\Gamma)$. Consequently

$$\int_{x_1}^{x_2} \sum_{i=1}^{\Phi(D,x,\Gamma)} \frac{1}{\cos \alpha_{i,x}(\Gamma)} dx = L_x(D,\Gamma), \tag{1.1.9}$$

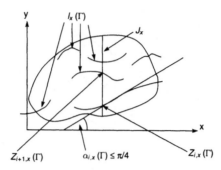

Figure 1.2

where $x_1 = \inf_{z \in \partial D} \operatorname{Re} z$ and $x_2 = \sup_{z \in \partial D} \operatorname{Re} z$. For a fixed y denote by J_y the set $D \cap \{z : \operatorname{Im} z = y\}$, by $\Phi(D, y, \Gamma)$ the number of points $z_{i,y}(\Gamma)$, where J_y intersects the union of arcs $l_y(\Gamma)$, and by $\alpha_{i,y}(\Gamma)$ the smaller angle between the axis y and the tangent to the arc $l_y(\Gamma)$ at the point $z_{i,y}(\Gamma)$. By similar arguments one can derive

$$\int_{y_1}^{y_2} \sum_{j=1}^{\Phi(D,y,\Gamma)} \frac{1}{\cos \alpha_{i,y}(\Gamma)} dx \, dy = L_y(D, \Gamma), \qquad (1.1.9')$$

where $y_1 = \inf_{z \in \partial D} \operatorname{Im} z$ and $y_2 = \sup_{z \in \partial D} \operatorname{Im} z$.

Now let $\Gamma = \Gamma(R)$. Denote $w(z_{i,x}(\Gamma)) = Re^{i\vartheta_{i,x}}$, also denote by $\beta_{i,x}(\Gamma(R))$ the smaller angle between the tangent to the arc $w(J_x)$ at the point $Re^{i\vartheta_{i,x}}$ and the ray $\{te^{i\vartheta_{i,x}}\}$. Observe that the length element of an arc $\Delta l(w(J_x))$ satisfies the relation

$$\Delta l(w(J_x)) \sim \frac{\Delta R}{\cos \beta_{i,x}(\Gamma(R))}$$

at the point $Re^{i\vartheta_{i,x}}$ (see Figure 1.3).

Hence,

$$\int_0^\infty \left(\sum_{i=1}^{\Phi(D,x,\Gamma(R))} \frac{1}{\Psi(R) \cos \beta_{i,x}(\Gamma(R))} \right) dR = \int_{J_x \cap D_x} \frac{|w'(z)|}{\Psi(|w(z)|)} dy \quad (1.1.10)$$

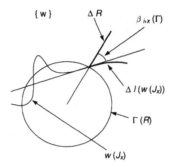

Figure 1.3

(as both sides of this equality represent complete masses on the curves $w(J_x \cap D_x)$ if these masses are distributed by the density $\Psi^{-1}(R) = \Psi^{-1}(|w(z)|)$).

Observe that due to the conformity, $\alpha_{i,x}(\Gamma(R)) = \beta_{i,x}(\Gamma(R))$, and the relation (1.1.10) is true in each point $z_{i,x}(\Gamma(R))$ (recall that multiple points were excluded from D). Consequently, the left-hand side integral in (1.1.10) is equal to

$$\int_0^\infty \left(\sum_{i=1}^{\Phi(D,x,\Gamma(R))} \frac{1}{\Psi(R)\cos\alpha_{i,x}(\Gamma(R))} \right) dR.$$

Thus, by relations (1.1.9) and (1.1.10)

$$\int_0^\infty \frac{L_x(D,\Gamma(R))}{\Psi(R)} dR$$

$$= \int_0^\infty \left(\int_{x_1}^{x_2} \left(\sum_{i=1}^{\Phi(D,x,\Gamma(R))} \frac{1}{\Psi(R)\cos\alpha_{i,x}(\Gamma(R))} \right) dx \right) dR$$

$$= \int_{x_1}^{x_2} \left(\int_0^\infty \left(\sum_{i=1}^{\Phi(D,x,\Gamma(R))} \frac{1}{\Psi(R)\cos\alpha_{i,x}(\Gamma(R))} \right) dR \right) dx$$

$$= \int_{x_1}^{x_2} \int_{J_x \cap D_x} \frac{|w'(z)|}{\Psi(|w(z)|)} dy\,dx = \int\int_{D_x} \frac{|w'(z)|}{\Psi(|w(z)|)} d\sigma.$$

Here the change of integration order is justified by Fubini's theorem. Repeating the arguments related with replacement of x by y, we obtain

$$\int_0^\infty \frac{L_y(D, \Gamma(R))}{\Psi(R)} dR = \int\int_{D_y} \frac{|w'(z)|}{\Psi(|w(z)|)} d\sigma.$$

Now (1.1.6) follows from (1.1.7), (1.1.8) and the last two identities.

1.1.4. Corollaries. First we shall extract a geometrical meaning from identity (1.1.6). Observe that

$$\int\int_{D_y} \frac{|w'(z)|}{\Psi(|w(z)|)} d\sigma = \int_{x_1}^{x_2} m_\Psi(J_x) dx = \int_{y_1}^{y_2} m_\Psi(J_y) dy,$$

where $m_\Psi(J_x)$ and $m_\Psi(J_y)$ are the complete masses distributed on the curves $w(J_x)$ and $w(J_y)$, respectively, by the density $\Psi^{-1}(|w(z)|)$. If $\Psi \equiv 1$, then $m_\Psi(J_x)$ and $m_\Psi(J_y)$ simply mean the lengths $l(w(J_x))$ and $l(w(J_y))$ of the curves $w(J_x)$ and $w(J_y)$, respectively. Therefore, the identity (1.1.6) can be rewritten in the forms

$$\int_0^\infty \frac{L(D, \Gamma(R))}{\Psi(R)} dR = \int_{x_1}^{x_2} m_\Psi(J_x) dx \qquad (1.1.11)$$

and

$$\int_0^\infty \frac{L(D, \Gamma(R))}{\Psi(R)} dR = \int_{y_1}^{y_2} m_\Psi(J_y) dy, \qquad (1.1.11')$$

and for $\Psi \equiv 1$ in the forms

$$\int_0^\infty L(D, \Gamma(R)) dR = \int_{x_1}^{x_2} l(w(J_x)) dx \qquad (1.1.11'')$$

and

$$\int_0^\infty L(D, \Gamma(R)) dR = \int_{y_1}^{y_2} l(w(J_y)) dy. \qquad (1.1.11''')$$

By virtue of their geometric meanings, the identities (1.1.11) and (1.1.11') can be named "principle of lengths and mass distributions", and the identities (1.1.11'') and (1.1.11''') "length principle".

Now we shall prove the inequalities in 1.1.2. The inequality (1.1.2) follows from (1.1.6) if we assume $\Psi \equiv 1$ and apply to the right-hand side of (1.1.6) the Cauchy–Buniakovski inequality. Assuming $\Psi \equiv 1 + R^2$ we similarly come to inequality (1.1.3). The inequality (1.1.2″) can be generalized as follows. Applying Hölder's inequality to the right-hand side of (1.1.6), for $p_1 > 0$, we obtain

$$\int_0^\infty \frac{L(D, \Gamma(R))}{\Psi(R)} dR \leq \left(\int \int_D \frac{|w'(z)|^{p_1}}{\Psi^{p_1}(|w(z)|)} d\sigma \right)^{1/p_1} (S(D))^{1-1/p_1}. \quad (1.1.12)$$

For proving (1.1.5) we shall use the following identity established by Hardy, Stein and Spencer (see Hayman [1], p. 56): for any $p > 0$

$$
\begin{aligned}
r \frac{d}{dr} J_p(r) &= \frac{p^2}{2\pi} \int_0^r \rho \int_0^{2\pi} |w'(\rho e^{i\vartheta})|^2 |w(\rho e^{i\vartheta})|^{p-2} d\vartheta d\rho \\
&= p^2 \int_0^\infty p(r, R) R^{p-1} dR.
\end{aligned}
$$

Assuming $\Psi(R) = R^{1-p/2}$ and $D = D(r)$, from (1.1.12) for $p_1 = 2$ and from the last identity we obtain

$$
\begin{aligned}
\int_0^\infty \frac{L(r, \Gamma(R))}{R^{1-p/2}} dR &\leq \left(\int_0^r \int_0^{2\pi} |w'(\rho e^{i\vartheta})|^2 |w(\rho e^{i\vartheta})|^{p-2} \rho d\rho d\vartheta \right)^{1/2} \pi^{1/2} r \\
&= \frac{\sqrt{2\pi} r^{3/2}}{p} \left[\frac{d}{dr} J_p(r) \right]^{1/2},
\end{aligned}
$$

whence (1.1.5) follows.

1.1.5. Estimates of integrals of lengths of pre-images of straight lines and rays. Let $\gamma(u)$ be the straight line $\{w : \operatorname{Re} w = u\}$, and let $\Psi(u)$, $u \in (-\infty, +\infty)$ be a continuous, positive function. Then the following identity holds:

$$\int_{-\infty}^{+\infty} \frac{L(D, \gamma(u))}{\Psi(u)} du = \int \int_D \frac{|w'(z)|}{\Psi(\operatorname{Re} w(z))} d\sigma. \quad (1.1.13)$$

One can prove this identity by repeating the arguments of the proof of (1.1.6) with the following replacements: R by u (also $\Gamma(R)$ by $\gamma(u)$, $\Psi(R)$ by $\Psi(u)$, ΔR by Δu) and $\beta_{i,x}(\Gamma(R))$ by the smaller angle between the tangent to the curve $w(J_x)$ at the point $w(z_{i,x}(\gamma(u)))$ and $\{w : \operatorname{Im} w = \operatorname{Im} w(z_{i,x}(\gamma(u)))\}$. Assuming $\Psi(u) \equiv 1$ and applying the Cauchy–Buniakovski inequality to

(1.1.13) we come to the following modification of the length–area principle in which the straight lines $\gamma(R)$ appear instead of circles $\Gamma(R)$:

$$\int_{-\infty}^{+\infty} L(D, \gamma(u))du \leq S_2^{1/2}(D)S^{1/2}(D). \qquad (1.1.14)$$

In the general case, from (1.1.13) and Hölder's inequality, the following similarity of (1.1.12) follows:

$$\int_{-\infty}^{+\infty} \frac{L(D, \gamma(u))}{\Psi(u)}du \leq \left(\int\int_D \frac{|w'(z)|^p}{\Psi^p(\operatorname{Re} w(z))}d\sigma\right)^{1/p} (S(D))^{1-1/p}. \qquad (1.1.15)$$

1.1.6. On magnitudes of integrals of lengths of ray pre-images. Let $\Gamma = \Gamma_\alpha = \{w : |w| > 0, \arg w = \alpha\}$, and let $\Psi(\alpha)$, $\alpha \in [0, 2\pi]$ be a positive, continuous function. Then the following identity holds:

$$\int_0^{2\pi} \frac{L(D, \Gamma(\alpha))}{\Psi(\alpha)}d\alpha = \int\int_D \left|\frac{w'(z)}{w(z)}\right| \frac{1}{\Psi(\arg w(z))}d\sigma. \qquad (1.1.16)$$

The proof of this identity, as well as the proof of (1.1.13), coincides with that of (1.1.6), where $\Gamma(R)$ is replaced by Γ_α and the other obvious changes are done.

Remark. The above modifications of the length–area principle and the main identity were given by the author, items 1.2.1–1.2.4, and by Sukiasian, items 1.2.5–1.2.6. They used a method for estimating Γ-lines developed in [4, 6, 9]; see also the next section. A similar identity exists, known as Coarea Formula in the Geometric Measure Theory, see Federer [1], Hardt [1], for much larger mappings $f : R^n \to R^m$, however, with the strong restriction that f is a Lipschitz function.

1.2. Tangent Variation Principle

1.2.1. In Section 1.1 it was mentioned that length–area principle, evaluating integrals of the length $L(\Gamma, D(R))$, is an efficient tool for investigation of various problems. It is natural to consider not only integrals of $L(D, \Gamma(R))$ but also *pure* estimates of $L(D, \Gamma(R))$ and, more generally, the estimates of $L(D, \Gamma)$, which is the total length of the pre-images $w^{-1}(\Gamma)$ for the given curves Γ. These pre-images we call Γ-lines. Similar estimates are of interest since the pre-images $w^{-1}(\gamma(u))$ become level sets of the function $\operatorname{Re} w(z)$ if the circle $\Gamma(R)$ is replaced by a straight line $\gamma(u)$. All the more, the physical meaning of these level sets is considered in almost any textbook of complex analysis, electro-hydrodynamics etc.

Nonetheless, a purely mathematical study of the Γ-lines and particularly study of their lengths started only recently,[1] probably, because of the absence of general methods of investigation of $w^{-1}(\Gamma)$.

The standard approach to evaluate $L(D,\Gamma)$ assumes estimation of the sum $\sum_i \int_{\Gamma_i} |z_i'(w)| dw$, where $z_i(w)$ are the branches of the function $z(w)$ inverse to $w(z)$, taken on the i-th sheet of the Riemann surface $F_{w(D)} = \{w(z) : z \in D\}$, and Γ_i are the sets in i-th sheet, projected to Γ. It was mentioned repeatedly in the literature that in general the problem of partitioning of a Riemann surface into sheets and separation of branches $z_i(w)$ is inaccessible. Such an approach can be used if the function w is a one-to-one mapping or if the structure of the Riemann surface $F_{w(D)}$ is known beforehand.

In this section we give a general method which we call "tangent variation principle". This principle permits to estimate the lengths $L(D,\Gamma)$ for wide classes of functions w and curves Γ.

1.2.2. Description of the method: The characteristic $V(D)$. In the remainder of this chapter, it will be assumed that D is a bounded domain with a piecewise smooth boundary and $w(z)$ is a function meromorphic in the closure of D.

We shall use the notation of Subsection 1.1.3. Observe that $\alpha_{i,x}(\Gamma) \leq \pi/4$ and $\alpha_{j,y}(\Gamma) < \pi/4$. Therefore it follows from the identities (1.1.7), (1.1.9) and (1.1.9') of Subsection 1.1.3 that

$$L(D,\Gamma) \leq \sqrt{2} \int_{x_1}^{x_2} \Phi(D,x,\Gamma)dx + \sqrt{2} \int_{y_1}^{y_2} \Phi(D,y,\Gamma)dy. \qquad (1.2.1)$$

---

[1] For the case of meromorphic functions and angular-quasiconformal mappings in $\mathbb{C}$ or in arbitrary complex domain from $\mathbb{C}$ and arbitrary smooth Jordan curves Γ see the author's papers [4, 6, 9]. Later on, the lengths of Γ-lines for particular class of one-to-one conformal mappings have been considered by Hayman and Wu [1], Garnett, Gehring and Jones [1], Fernandez and Hamilton [1], Fernandez and Zinsmeister [1], Fernandez, Heinonen and Martio [1], Fernandez [1], Bishop, Carleson, Garnett and Jons [1], Bishop and Jons [1], Väisälä [1], Öyma [1], Astala, Fernandez and Rohde [1]; for quasiregular one-to-one mappings see Väisälä [1].

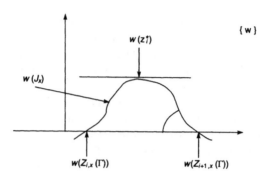

Figure 1.4

For a given x, let $\{m_x^{(p)}\}$ denote the collection of interval components of the set J_x, let $m_{x,i}^{(p)} \subset m_x^{(p)}$ be the interval with the endpoints $z_{i,x}(\Gamma)$ and $z_{i+1,x}(\Gamma)$, and let Φ_p be the number of points $z_{i,x}(\Gamma)$ on $m_x^{(p)}$.

Assume that Γ is the real axis and for a given x all singular points of $w(z)$ avoid J_x. Also suppose that $\Phi_p \geq 2$. The endpoints $w(z_{i,x}(\Gamma))$ and $w(z_{i+1,x}(\Gamma))$ of the curve $w(m_{x,i}^{(p)})$ are on the real axis by the definition of $z_{i,x}(\Gamma)$. The smaller angle between the tangent to this curve and the real axis is greater than $\pi/4$ in each of these endpoints. This follows from the conformity of $w(z)$ in that points. Evidently, there exists a point z_i^* between $z_{i,x}(\Gamma)$ and $z_{i+1,x}(\Gamma)$, such that the tangent to $w(m_{x,i}^{(p)})$ at $w(z_i^*)$ is parallel to the real axis. Therefore, the total variation of the slope of the tangent to $w(m_{x,i}^{(p)})$, while the point $w(z_{i,x}(\Gamma))$ moves to the point $w(z_i^*)$ (or $w(z_i^*)$ moves to $w(z_{i+1,x}(\Gamma)))$, is greater than $\pi/4$ (see Figure 1.4).

So, denoting by $\alpha_x(y) = \arg(\partial/\partial y)w(z)$ the angle between the real axis and the tangent to $w(J_x)$ at $w(z)$, $z = x + iy$, we get

$$Var_{m_{x,i}^{(p)}}\alpha_x(y) > \frac{\pi}{2},\qquad(1.2.2)$$

where $Var_X f$ is the variation of a function f on X. Summing up over all i gives

$$Var_{m^{(p)}_{x,i}} \alpha_x(y) \geq (\Phi_p - 1) \frac{\pi}{2}$$

or

$$\Phi_p \leq \frac{2}{\pi} Var_{m^{(p)}_{x,i}} \alpha_x(y) + 1. \tag{1.2.3}$$

Obviously, the last inequality remains valid also for $\Phi_p < 2$. Summing up (1.2.3) over all p we get

$$\Phi(D, x, \Gamma) = \sum_{\{m^{(p)}_x\}} \Phi_p \leq \frac{2}{\pi} \sum_{\{m^{(p)}_x\}} Var_{m^{(p)}_x} \alpha_x(y) + \sum_{\{m^{(p)}_x\}} 1$$

$$= \frac{2}{\pi} Var_{J_x} \alpha_x(y) + \sum_{\{m^{(p)}_x\}} 1. \tag{1.2.4}$$

Suppose that the part of the domain D contained in the strip between lines $\{z : \operatorname{Re} z = x'\}$ and $\{z : \operatorname{Re} z = x''\}$ is decomposed into n connected components with boundaries having common points both with $J_{x'}$ and $J_{x''}$. Then for every $x \in (x', x'')$

$$\sum_{\{m^{(p)}_x\}} 1 = n,$$

where the constant n is independent of x and consequently

$$\int_{x'}^{x''} \left(\sum_{\{m^{(p)}_x\}} 1 \right) dx = n(x'' - x').$$

But the quantity $n(x'' - x')$ is the sum of projections of the above mentioned components on X-axis. It is easy to see that $2n(x'' - x')$ is less than the length of the part of ∂D contained in the mentioned strip. Hence splitting the interval (x_1, x_2) into appropriate parts we obtain

$$\int_{x_1}^{x_2} \left(\sum_{\{m^{(p)}_x\}} 1 \right) dx \leq \frac{l(D)}{2}, \tag{1.2.5}$$

where $l(D)$ is the length of the boundary ∂D of D. In view of this, integration of (1.2.4) gives

$$\int_{x_1}^{x_2} \Phi(D,x,\Gamma)dx < \frac{2}{\pi}\int_{x_1}^{x_2} Var_{J_x}\alpha_x(y)dx + \frac{l(D)}{2}. \qquad (1.2.6)$$

So, we evaluated $\Phi(D,x,\Gamma)$ for those points x for which the singularities of w avoid J_x. As there is only a finite number of such singular points x, they do not affect the value of the integral.

Similar to the above arguments. replacing x by y and denoting by $a_y(x) = \arg\frac{\partial}{\partial x}w(z)$ the angle between the tangent to $w(J_y)$ at $w(z)$ and the real axis, we derive

$$\int_{y_1}^{y_2} \Phi(D,y,\Gamma)dy < \frac{2}{\pi}\int_{y_1}^{y_2} Var_{J_y}\alpha_y(x)dy + \frac{l(D)}{2}. \qquad (1.2.7)$$

Summing up the inequalities (1.2.6) and (1.2.7) and taking into account (1.2.1), we find

$$L(D,\Gamma) < \frac{2\sqrt{2}}{\pi}\left\{\int_{x_1}^{x_2} Var_{J_x}\alpha_x(y)dx + \int_{y_1}^{y_2} Var_{J_y}\alpha_y(x)dy\right\} + \sqrt{2}l(D).$$
$$(1.2.8)$$

This inequality is the Tangent Variation Principle in the case where Γ is the real axis. The expression in the figure brackets is denoted by $V(D)$. Evidently $V(D)$ takes the role of the characteristic function while estimating $L(D,\Gamma)$.

Obviously

$$\begin{aligned}
V(D) &= \int_{x_1}^{x_2} Var_{J_x}\alpha_x(y)dx + \int_{y_1}^{y_2} Var_{J_y}\alpha_y(x)dy \\
&= \int_{x_1}^{x_2}\left(\int_{J_x}\left|\frac{\partial}{\partial y}\arg\frac{\partial}{\partial y}w(z)\right|dy\right)dx \\
&\quad + \int_{y_1}^{y_2}\left(\int_{J_y}\left|\frac{\partial}{\partial x}\arg\frac{\partial}{\partial x}w(z)\right|dx\right)dy \\
&\leq \int\int_D\left|\frac{\partial}{\partial x}\ln|w'(z)|\right|d\sigma + \int\int_D\left|\frac{\partial}{\partial y}\ln|w'(z)|\right|d\sigma \\
&\leq 2B(r), \qquad\qquad\qquad\qquad\qquad\qquad\qquad (1.2.9)
\end{aligned}$$

where

$$B(r) := \int \int_D \left| \frac{w''(z)}{w'(z)} \right| d\sigma.$$

Thus, denoting $V(x, D) := Var_{J_x} \alpha_x(y)$; $V(y, D) := Var_{J_y} |a_y(x)$, we find

$$V(D) = \int_{x_1}^{x_2} V(x, D) dx + \int_{y_1}^{y_2} V(y, D) dy. \qquad (1.2.10)$$

1.2.3. The general case. Assuming that Γ is a smooth Jordan curve, consider the magnitude

$$\nu(\Gamma) = Var_{z \in \Gamma} \alpha_\Gamma(z), \qquad (1.2.11)$$

where $\alpha_\Gamma(z)$ is the angle between the tangent to Γ at $z \in \Gamma$ and the real axis. Evidently, $\nu(\Gamma)$ in a sense characterizes the "curvature" of Γ.

First we shall establish an inequality similar to (1.2.8) for curves of small curvature $\nu(\Gamma) < \pi/4$. Let $\alpha_i^*(+)$ be the smaller angle between the tangent to $w(J_x)$ at $w(z_{i,x}(\Gamma))$ and the line joining $w(z_{i,x}(\Gamma))$ and $w(z_{i+1,x}(\Gamma))$, let $\widetilde{\alpha}_i^*(+)$ be the smaller angle between the tangent to Γ at $w(z_{i,x}(\Gamma))$ and the line joining $w(z_{i,x}(\Gamma))$ and $w(z_{i+1,x}(\Gamma))$ and let $\widetilde{\alpha}_i(+)$ be the smaller angle between the curves $w(J_x)$ and Γ at $w(z_{i,x}(\Gamma))$. Interchanging the points $z_{i,x}(\Gamma)$ and $z_{i+1,x}(\Gamma)$ we similarly define the angles $\alpha_{i+1}^*(-)$, $\widetilde{\alpha}_{i+1}^*(-)$ and $\widetilde{\alpha}_{i+1}(-)$. Suppose that the set J_x does not contain singular points of $w(z)$. Then there is a point $z_i^* \in m_{x,i}^{(p)}$ such that the tangent to the curve $w(m_{x,i}^{(p)})$ at $w(z_i^*)$ is parallel to the line joining $w(z_{i,x}(\Gamma))$ and $w(z_{i+1,x}(\Gamma))$. Therefore, the complete variation of slope of the tangent to the curve $w(m_{x,i}^{(p)})$ while moving from the point $w(z_{i,x}(\Gamma))$ to the point $w(z_i^*)$ (or while moving from $w(z_i^*)$ to $w(z_{i+1,x}(\Gamma))$) is not less than $\alpha_i^*(+)$ (or $\alpha_{i+1}^*(-)$). Hence

$$Var_{\{m_{x,i}^{(p)}\}} \alpha_x(y) \geq \alpha_i^*(+) + \alpha_{i+1}^*(-). \qquad (1.2.12)$$

It follows from the definition of the angles $\alpha_i^*(+)$, $\widetilde{\alpha}_i^*(+)$ and $\widetilde{\alpha}_i(+)$ that the following three cases are possible:

$$\begin{aligned}
\alpha_i^*(+) &= \widetilde{\alpha}_i(+) - \widetilde{\alpha}_i^*(+), \\
\alpha_i^*(+) &= \widetilde{\alpha}_i(+) + \widetilde{\alpha}_i^*(+), \\
\alpha_i^*(+) &= \widetilde{\alpha}_i^*(+) - \widetilde{\alpha}_i(+),
\end{aligned}$$

see respectively Figures 1.5, 1.6 and 1.7.

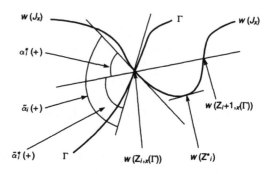

Figure 1.5

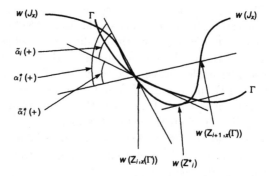

Figure 1.6

Hence, in any case

$$\alpha_i^*(+) + \alpha_{i+1}^*(-) > \widetilde{\alpha}_i(+) + \widetilde{\alpha}_{i+1}(-) - [\widetilde{\alpha}_i^*(+) + \widetilde{\alpha}_{i+1}^*(-)].$$

Since $\widetilde{\alpha}_i^*(+) + \widetilde{\alpha}_{i+1}^*(-) < \nu(\Gamma)$, from (1.2.12) it follows that

$$Var_{\{m_{x,i}^{(p)}\}}\alpha_x(y) \geq \widetilde{\alpha}_i(+) + \widetilde{\alpha}_{i+1}(-) - \nu(\Gamma). \qquad (1.2.13)$$

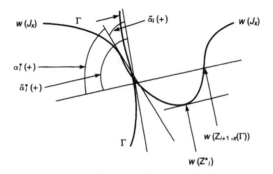

Figure 1.7

Note that $\widetilde{\alpha}_i(+)$ is the smaller angle between the curves Γ and $w(J_x)$ at the point $w(z_{i,x}(\Gamma))$. By the definition of points $z_{i,x}(\Gamma)$ and by conformity of the mapping $w(z)$ in the points of J_x $\widetilde{\alpha}_i(+) = \pi/2 - \alpha_{i,x}(\Gamma)$ and $\alpha_{i,x}(\Gamma) \leq \pi/4$. Hence $\widetilde{\alpha}_i(+) \geq \pi/4$, and similarly $\widetilde{\alpha}_{i+1}(-) \geq \pi/4$. Thus, the following similarity of the inequality (1.2.2) is true:

$$Var_{\{m_{x,i}^{(p)}\}}\alpha_x(y) \geq \frac{\pi}{2} - \nu(\Gamma). \tag{1.2.14}$$

By arguments similar to those used for deriving the inequality (1.2.8), from (1.2.2) we get

$$\Phi_p \leq \frac{1}{2^{-1}\pi - \nu(\Gamma)}Var_{\{m_x^{(p)}\}}\alpha_x(y) + 1.$$

Summation over all p gives

$$\Phi(D, x, \Gamma) \leq \frac{1}{2^{-1}\pi - \nu(\Gamma)}Var_{\{J_x\}}\alpha_x(y) + \sum_{\{m_x^{(p)}\}} 1.$$

Integrating this inequality and taking into account (1.2.5) and the finiteness of the singular points of $w(z)$ in D, we obtain

$$\int_{x_1}^{x_2} \Phi(D, x, \Gamma)dx \leq \frac{1}{2^{-1}\pi - \nu(\Gamma)}V(x, D) + \frac{l(D)}{2},$$

and similarly

$$\int_{y_1}^{y_2} \Phi(D, y, \Gamma) \leq \frac{1}{2^{-1}\pi - \nu(\Gamma)} V(y, D) + \frac{l(D)}{2}.$$

The last two estimates and (1.2.1) imply the inequality

$$L(D, \Gamma) \leq \frac{\sqrt{2}}{2^{-1}\pi - \nu(\Gamma)} V(D) + \sqrt{2}l(D), \qquad (1.2.15)$$

being the Tangent Variation Principle for the case $\nu(\Gamma) < \pi/4$.

Let now $\varepsilon_0 = \text{const} < \pi/4$ and let Γ be any smooth, bounded or unbounded Jordan curve with $\varepsilon_0 \leq \nu(\Gamma) < \infty$.

Starting from one of the endpoints of Γ, we successively split Γ into parts $\Gamma^{(\nu)}$, $\nu = 1, \ldots, k$, such that $\Gamma^{(\nu)} = \varepsilon_0$ for any ν. If $\bigcup_{\nu=1}^{k} \Gamma^{(\nu)} \neq \Gamma$, then $\nu(\Gamma^{(k+1)}) < \varepsilon_0$, where $\Gamma^{(k+1)} = \Gamma \setminus \left[\bigcup_{\nu=1}^{k} \Gamma^{(\nu)} \right]$. Therefore, the total number k (or $k+1$) of such parts at most is equal to $[\nu(\Gamma)/\varepsilon_0] + 1$, where $[x]$ is the integer part of x, and $\nu(\Gamma^{(\nu)}) \leq \varepsilon_0 < \frac{\pi}{4}$ for any $\Gamma^{(\nu)}$.

Applying the inequality (1.2.15) to all $\Gamma^{(\nu)}$ we obtain

$$\begin{aligned}
L(D, \Gamma) &= \sum_{(\nu)} L(D, \Gamma^{(\nu)}) \\
&\leq \sqrt{2} \left(\left[\frac{\nu(\Gamma)}{\varepsilon_0} \right] + 1 \right) \left(\frac{\pi}{2} - \varepsilon_0 \right)^{-1} V(D) \\
&\quad + \sqrt{2} \left(\left[\frac{\nu(\Gamma)}{\varepsilon_0} \right] + 1 \right) l(D).
\end{aligned}$$

Hence, choosing ε_0 sufficiently close to $\pi/4$ we get

$$L(D, \Gamma) < 3 \left(\nu(\Gamma) + 1 \right) \left(V(D) + l(D) \right).$$

Denoting $K(\Gamma) = 3(\nu(\Gamma) + 1)$ we come to the following result.

The First Fundamental Theorem (Tangent Variation Principle).
For any meromorphic function $w(z)$ in $\bar{D}$ and for any smooth Jordan curve Γ

$$L(D, \Gamma) \leq K(\Gamma)(V(D) + l(D)). \qquad (1.2.16)$$

Thus, the length $L(D(\Gamma))$ is estimated by the magnitudes $\nu(\Gamma)$ and $V(D)$, both indicating the "curvature" (or which is the same as the tangent variations) of the curves Γ, $w(J_x)$ and $w(J_y)$.

1.3. Estimates for Collections of Γ-Lines

1.3.1. Below, we shall consider finite collections of disjoint bounded smooth Jordan curves $\Gamma_1, \ldots, \Gamma_q$. We shall denote by K the different absolute constants and by $h = h(\Gamma_1, \ldots, \Gamma_q, \alpha)$ the constants depending only on $\Gamma_1, \ldots, \Gamma_q$ and α.

Applying Theorem 1 to a collection $\Gamma_1, \ldots, \Gamma_q$, we come to the inequality

$$\sum_{\nu=1}^{q} L(D, \Gamma_\nu) \leq \left(\sum_{\nu=1}^{q} K(\Gamma_\nu) \right) (V(D) + l(D)).$$

In Section 4.1 it will be shown that $V(D(r)) \leq KrA(r)$ as $r = r_n \to 0$ when $D = D(r)$ and $w(z)$ is meromorphic in $\mathbb{C}$. Therefore

$$\sum_{\nu=1}^{q} L(r, \Gamma_n u) \leq K \left(\sum_{\nu=1}^{q} K(\Gamma_\nu) \right) rA(r) + 2\pi r \sum_{\nu=1}^{q} K(\Gamma_\nu).$$

However, we can prove statements like deficiency relations for curves only if the coefficient of $r \cdot A(r)$ is an absolute constant. For this, we shall establish now the relation

$$\sum_{\nu=1}^{q} L(r, \Gamma_\nu) \leq KV(D_r) + S(w, D_r, \Gamma_1, \ldots, \Gamma_q),$$

where $S = o(V(D_r))$ or which is the same as $S = o(rAr)$ as $r = r_n \to \infty$.

In other words, we need an analog of the Principle for Γ-lines which in a sense will take the same role in the theory of Γ-lines as that of the Second Fundamental Theorem in Nevanlinna theory.

1.3.2. The main theorem. For the collection of Γ-lines of arbitrary functions in D it is given in the following:

Second Fundamental Theorem. *Let $w(z)$ be a function meromorphic in the closure of D and let Γ_ν, $\nu = 1, 2, \ldots, q$, be some disjoint, bounded smooth Jordan curves with $\nu(\Gamma_\nu) < \infty$. Then,*

$$\sum_{i=1}^{q} L(D, \Gamma_\nu) \leq KV(D_r) + h(\Gamma_1, \ldots, \Gamma_q) A_1(D) + \sqrt{2} l(D). \qquad (1.3.1)$$

Further, if Γ is an unbounded smooth Jordan curve with $\nu(\Gamma) < \infty$, then

$$L(D,\Gamma) \le KV(D) + h(\Gamma)A_1(D) + 3\sqrt{2}l(D). \qquad (1.3.1')$$

The proof of the inequality (1.3.1) is based on the following

Lemma 1. *Let w and $\Gamma_1, \Gamma_2, \ldots, \Gamma_q$ be the same as in the above theorem. Assume that $\mu = \mu(t)$, $t \in (0,1)$, is a twice continuously differentiable curve entirely contained in $\bar{D}$, $L(\mu)$ is the length of the curve $w(\mu(t))$ ($t \in (0,1)$) in the spherical metric, $n^*(\mu)$ is the number of singular points $z_i \in \mu$ of the function w, and $\Phi_\alpha(\mu, \Gamma_\nu)$ is the number of points $z_i \in \mu$ for which*
(i) *$w(z_i) \in \Gamma$ and*
(ii) *the smaller angle between the curves $w(\mu(t))$ and Γ_ν at $w(z_i)$ is not less than $\alpha > 0$.*
Then

$$\sum_{\nu=1}^{q} \Phi_\alpha(\mu, \Gamma_\nu) \le \frac{2}{\alpha} \int_0^1 \left| \frac{\partial}{\partial t} \arg \frac{\partial}{\partial t} w(\mu(t)) \right| dt$$
$$+ h(\Gamma_1, \ldots, \Gamma_q, \alpha) L(\mu) + 2n^*(\mu) + 1, \qquad (1.3.2)$$

and if Γ is an unbounded smooth Jordan curve with $\nu(\Gamma) < \infty$, then

$$\Phi_\alpha(\mu, \Gamma) \le \frac{6}{\alpha} \int_0^1 \left| \frac{\partial}{\partial t} \arg \frac{\partial}{\partial t} w(\mu(t)) \right| dt$$
$$+ h(\Gamma, \alpha) L(\mu) + 6n^*(\mu) + 3, \qquad (1.3.2')$$

Proof. Let the points z_i be the same as in the definition of $\Phi_\alpha(\mu, \Gamma_\nu)$, and suppose that they are numerated for all ν, $\nu = 1, \ldots, q$. First we shall prove (1.3.2) under the assumption that the number of the points z_i is not less than 2. We remunerate the points z_i on $\mu(t)$ in the order of increasing t, from 0 to 1, and split them into the following subsets:

(a) Denote by $z_i' \in \{z_i\}$ those points z_i for which for at least one of the points z_{i-1} and z_{i+1} and the points $w(z_{i-1})$ and $w(z_i)$ (respectively, $w(z_i)$ and $w(z_{i+1})$) lie on different curves Γ_ν, $\nu = 1, \ldots, q$. Denote by n' the number of such points and by t_i' the numbers for which $\mu(t_i') = z_i'$.

(b) Denote by $z_i^j \in \{z_i\}$ the points that lie between z_j' and z_{j+1}' after the renumeration of z_i, by n_j the number of these points, and by t_i^j the points for which $\mu(t_i^j) = z_i^j$.

Evidently,

$$\sum_{\nu=1}^{q} \Phi_{\alpha}(\mu, \Gamma_{\nu}) = n' + \sum_{\{z'_j\}} n_j. \qquad (1.3.3)$$

First we show that

$$n' < h(\Gamma_1, \ldots, \Gamma_q)L(\mu). \qquad (1.3.4)$$

Let R be a fixed number such that the curves Γ_{ν}, $\nu = 1, \ldots, q$, lie in the disk $|w| \leq R/2$, and let C_0 be the minimal distance between the curves Γ_{ν}, $\nu = 1, \ldots, q$. Then for every j, such that $w(z'_j)$ and $w(z'_{j+1})$ lie on different curves Γ_{ν}, $w(\mu(t))$ maps the intercept $[t'_j, t'_{j+1}]$ to a curve l_j with endpoints on different Γ_{ν}, and the length $|l_j|$ of the curve or the set of curves $|l_j \cap \{w : |w| < R\}|$ is not less than C_0. Hence

$$n' \leq \frac{2}{C_0} \sum_{\{z'_j\}} |l_j|. \qquad (1.3.5)$$

Since $|l_i| \leq (1 + R^2)L_i$, where L_i is the spherical length of the curve l_i, from (1.3.5) we obtain (1.3.4) since

$$n' \leq \frac{2(1 + R^2)}{C_0} \sum_{\{z'_j\}} L_j \leq \frac{2(1 + R^2)}{C_0} L(\mu) = h(\Gamma_1, \ldots, \Gamma_q)L(\mu).$$

Now we shall show that for any j

$$n_j \leq \frac{2}{\alpha} \int_{t'_j}^{t'_{j+1}} \left| \frac{\partial}{\partial t} \arg \frac{\partial}{\partial t} w(\mu(t)) \right| dt + hL_j + 2n_j^*, \qquad (1.3.6)$$

where $h < \infty$ is a constant depending only on $\Gamma_1, \ldots, \Gamma_q$ and α, and n_j^* is the number of singular points of $w(z)$ on l_j.

It is clear that all z_i^j belong to a curve Γ_{ν}. We split Γ_{ν} into finite number of parts Γ_{ν}^k so that the variation of angle between the tangent to Γ_{ν} and the real axis be less than $\alpha/4$ on each adjacent triad Γ_{ν}^k, Γ_{ν}^{k+1} and Γ_{ν}^{k+2}. Let h_{ν}^k be the minimal distance between Γ_{ν}^k and $\Gamma_{\nu}^{k'}$ for $k' \neq k - 1, k + 1$, and let $c_{\nu} = \min_k h_{\nu}^k$. Constructing the corresponding partitions for all Γ_{ν}, $\nu = 1, \ldots, q$, we set $c' = \min_{\nu} c_{\nu}$.

Now the following cases (A), (B), (C) are possible:

(A) The pair of points z'_j, z^j_1 (z^j_i, z^j_{i+1} or $z^j_{n_j}$, z'_{j+1} respectively) lie either on a single curve Γ^k_ν or on two adjacent curves Γ^k_ν, and for $t \in [t'_j, t^j_1]$ ($t \in [t^j_i, t^j_{i+1}]$ and $t \in [t^j_{n_j}, t'_{j+1}]$ respectively) the curve $w(\mu(t))$ is free from singularities of $w(z)$, and n'_j denotes the number of such pairs.

It is easy to see that between t'_j and t^j_1 there exists a point t^* such that the tangent to the curve $w(\mu(t))$ at $z^* = \mu(t^*)$ is parallel to the line joining $w(z'_j)$ and $w(z^j_1)$. The angles between this line and the curve $w(\mu(t))$ at the points $w(z'_j)$ and $w(z'_1)$ are not less than $\alpha/2$. Therefore, it is evident that the absolute value of the difference of the angles between the tangents to $w(\mu(t))$ at $w(z'_j)$ and $w(z^*)$ (at $w(z^*)$ and $w(z^j_1)$ respectively) is not less than $\alpha/2$. Since the angle between the tangent to $w(\mu(t))$ and the real axis is $\arg \frac{\partial}{\partial t} w(\mu(t))$, we obtain

$$\int_{t'_j}^{t^*} \left| \frac{\partial}{\partial t} \arg \frac{\partial}{\partial t} w(\mu(t)) \right| dt \geq \frac{\alpha}{2} \quad \text{and} \quad \int_{t^*}^{t^j_1} \left| \frac{\partial}{\partial t} \arg \frac{\partial}{\partial t} w(\mu(t)) \right| dt \geq \frac{\alpha}{2}.$$

Hence

$$\int_{t'_j}^{t^j_1} \left| \frac{\partial}{\partial t} \arg \frac{\partial}{\partial t} w(\mu(t)) \right| dt \geq \alpha.$$

Similarly, for the pairs z^j_i, z^j_{i+1} or $z^j_{n_j}$, z'_{j+1} satisfying condition (A)

$$\int_{t^j_i}^{t^j_{i+1}} \left| \frac{\partial}{\partial t} \arg \frac{\partial}{\partial t} w(\mu(t)) \right| dt \geq \alpha \quad \text{and} \quad \int_{t^j_{n_j}}^{t'_{j+1}} \left| \frac{\partial}{\partial t} \arg \frac{\partial}{\partial t} w(\mu(t)) \right| dt \geq \alpha.$$

Summing up these estimates we get

$$n'_j \leq \frac{1}{\alpha} \int_{t'_j}^{t'_{j+1}} \left| \frac{\partial}{\partial t} \arg \frac{\partial}{\partial t} w(\mu(t)) \right| dt. \tag{1.3.7}$$

(B) The pair of points z'_j, z^j_1 (z^j_i, z^j_{i+1} or $z^j_{n_j}$, z'_{j+1} respectively) lie on different curves Γ^k_ν and $\Gamma^{k'}_\nu$, where $k' \neq k - 1$, $k' \neq k + 1$, and the curve $w(\mu(t))$ does not contain singularities of $w(z)$ for $t \in [t'_j, t^j_1]$ ($t \in [t^j_i, t^j_{i+1}]$ and $t \in [t^j_{n_j}, t'_{j+1}]$ respectively), and n''_j is the number of such pairs.

Denote by $|l(z'_j, z^j_1)|$, $|l(z^j_i, z^j_{i+1})|$ and $|l(z^j_{n_j}, z'_{j+1})|$ the lengths of the parts of the curve $w(\mu(t)) \cap \{w : |w| \leq R\}$ corresponding to the segments $[t'_j, t^j_1]$,

$[t_i^j, t_{i+1}^j]$ and $[t_{n_j}^j, t_{j+1}']$, respectively. In case (B) each of these lengths is not less than c'. On the other hand, the sum of all these lengths is at most $|l_j|$. Therefore, $n_j'' c' \leq |l_j|$ and

$$n_j'' \leq \frac{1}{c'}|l_j| \leq \frac{1+R^2}{c'}L_j, \tag{1.3.8}$$

$$\sum_{(j)} n_j'' \leq \frac{1+R^2}{c'} \sum_{(j)} L_j \leq h(\Gamma_1, \ldots, \Gamma_q)L(\mu). \tag{1.3.8'}$$

(C) The pairs z_j', z_1^j (z_i^j, z_{i+1}^j or $z_{n_j}^j$, z_{j+1}' respectively) for which at least one singular point of $w(z)$ lies on the part of the curve $w(\mu(t))$ corresponding to the segment $[t_j', t_1^j]$ ($[t_i^j, t_{i+1}^j]$ or $[t_{n_j}^j, t_{j+1}']$ respectively), and n_j''' is the number of such pairs. In this case, obviously

$$n_j''' \leq n_j^*, \tag{1.3.9}$$

$$\sum_{(j)} n_j''' \leq n^*(\mu). \tag{1.3.9'}$$

The inequality (1.3.6) is a consequence of (1.3.7), (1.3.8) and (1.3.9) and of the obvious relation $n_j \leq 2(n_j' + n_j'' + n_j''')$. Therefore, it follows from the relations (1.3.3), (1.3.4), (1.3.6), (1.3.8') and (1.3.9') that

$$\sum_{\nu=1}^{q} \Phi_\alpha(\mu, \Gamma_\nu) \leq \frac{2}{\alpha} \int_0^1 \left| \frac{\partial}{\partial t} \arg \frac{\partial}{\partial t} w(\mu(t)) \right| dt + hL(\mu) + 2n_\mu^*,$$

if we assume that the number of z_i points is not less than 2. If such a point z_i is unique, then we merely add 1 to the right-hand side of the last inequality. This completes the proof of (1.3.2).

Now let us prove estimate (1.3.2'). As $\nu(\Gamma) < \infty$ for the considered Jordan curve Γ, it can be split into two connected components Γ_1 and Γ_2 (or three connected components Γ_1, Γ_2 and Γ_3) so that Γ_1 is bounded and $\nu(\Gamma_2) \leq \alpha/4$ (or $\nu(\Gamma_2), \nu(\Gamma_3) \leq \alpha/4$). We apply the inequality (1.3.2) of Lemma 1 to the curve Γ_1. Also we apply the same lemma to the curves Γ_2 and Γ_3. Taking into account that for Γ_2 and Γ_3 only cases (A) and (C) are possible, we obtain

$$\Phi_\alpha(\mu, \Gamma_\nu) \leq \frac{2}{\alpha} \int_0^1 \left| \frac{\partial}{\partial t} \arg \frac{\partial}{\partial t} w(\mu(t)) \right| dt + 2n_\mu^* + 1, \quad \nu = 2, 3.$$

Summing up the estimates obtained for Γ_1, Γ_2 and Γ_3 we come to (1.3.2′).

For proving the inequality (1.3.1) we suppose that no singular points of $w(z)$ lie on J_x for a given x. Then we apply Lemma 1 with $\alpha = \pi/4$, $\mu = \mu^{(p)} = m_x^{(p)}$.

Observe that by the definition of points $z_{i,x}(\Gamma)$ and conformity of w in the points of J_x

$$\Phi(D, x, \Gamma_\nu) = \sum_{(p)} \Phi_{\pi/4}(m_x^{(p)}, \Gamma_\nu), \qquad (1.3.10)$$

$$\sum_{(p)} \int_0^1 \left| \frac{\partial}{\partial t} \arg \frac{\partial}{\partial t} w(\mu(t)) \right| dt = \sum_{(p)} Var_{m_x^{(p)}} \alpha_x(y) = Var_{J_x} \alpha_x(y), \quad (1.3.11)$$

$$\sum_{(p)} L_{m_x^{(p)}} = \int_{J_x} \frac{|w'(z)|}{1 + |w(z)|^2} dy, \qquad (1.3.12)$$

$$\sum_{(p)} n^*_{m_x^{(p)}} = 0. \qquad (1.3.13)$$

Therefore, from the inequality (1.3.2) of Lemma 1 it follows that

$$\sum_{\nu=1}^q \Phi(D, x, \Gamma_\nu) \leq \frac{8}{\pi} Var_{J_x} \alpha_x(y) + h \int_{J_x} \frac{|w'(z)|}{1 + |w(z)|^2} dy + \sum_{\{m_x^{(p)}\}} 1.$$

As the function w has a finite set of singular points, integrating the last estimate by x from x_1 to x_2 and using the inequality (1.2.5) of Section 1.2 we get

$$\sum_{\nu=1}^q \int_{x_1}^{x_2} \Phi(D, x, \Gamma_\nu) dx$$

$$\leq \frac{8}{\pi} \int_{x_1}^{x_2} Var_{J_x} \alpha_x(y) dx + h \iint_D \frac{|w'(z)|}{1 + |w(z)|^2} d\sigma + \frac{l(D)}{2}.$$

Similarly, we get

$$\sum_{\nu=1}^q \int_{y_1}^{y_2} \Phi(D, y, \Gamma_\nu) dy$$

$$\leq \frac{8}{\pi} \int_{y_1}^{y_2} Var_{J_y} \alpha_y(x) dy + h \iint_D \frac{|w'(z)|}{1 + |w(z)|^2} d\sigma + \frac{l(D)}{2}.$$

Now estimate (1.3.1) follows from (1.2.1) of Section 1.2 and the last two inequalities. The proof of inequality (1.3.1′) is similar, the only difference is the use of (1.3.2′) instead of (1.3.2).

1.3.3. An improvement. How much information we have to use for describing the lengths of Γ-lines? The next theorem shows that if we only know the behavior of our function in small neighborhoods of Γ-lines, then we can describe these lengths. This is important for applications, since it means that in a sense the lengths of the Γ-lines depend only on the behavior of w in the mentioned small neighborhoods.

Improvement of the second fundamental theorem. *Let the conditions of the previous theorem be satisfied. Further, let $\varepsilon = \text{const} > 0$, $\Gamma_\nu(\varepsilon)$, $\nu = 1, 2, \ldots, q$, be some ε-neighborhoods of the curves Γ_ν and let*

$$J_x^{(\varepsilon)} = \left(\bigcup_{\nu=1}^{q} w^{-1}(\Gamma_\nu(\varepsilon)) \right) \cap J_x, \quad J_y^{(\varepsilon)} = \left(\bigcup_{\nu=1}^{q} w^{-1}(\Gamma_\nu(\varepsilon)) \right) \cap J_y,$$

$$D_\varepsilon = D \cap \bigcup_{\nu=1}^{q} w^{-1}(\Gamma_\nu(\varepsilon)).$$

Then

$$\sum_{\nu=1}^{q} L(D, \Gamma_\nu) \leq K \left[\int_{x_1}^{x_2} Var_{J_x^{(\varepsilon)}} \alpha_x(y) dx + \int_{y_1}^{y_2} Var_{J_y^{(\varepsilon)}} \alpha_y(x) dy \right]$$

$$+ h(\Gamma_1, \ldots, \Gamma_q, \varepsilon) \left[\int_{x_1}^{x_2} \int_{J_x^{(\varepsilon)}} \frac{|w'(z)|}{1 + |w(z)|^2} dx dy \right.$$

$$\left. + \int_{y_1}^{y_2} \int_{J_y^{(\varepsilon)}} \frac{|w'(z)|}{1 + |w(z)|^2} dx dy \right] + \sqrt{2} l(D)$$

$$\leq 2K \int \int_{D_\varepsilon} \left| \frac{w''(z)}{w'(z)} \right| d\sigma + h(\Gamma_1, \ldots, \Gamma_q, \varepsilon)$$

$$\times \int \int_{D_\varepsilon} |w(z)| d\sigma + \sqrt{2} l(D). \qquad (1.3.14)$$

This theorem means that estimates of lengths of Γ_ν-lines can be obtained on the basis of certain information about the behavior of $w(z)$ on the pre-images of small neighborhoods of the curves Γ_ν.

Proof of the theorem. Obviously

$$\Phi(D, x, \Gamma_\nu) = \sum_{(p)} \Phi_{\pi/4}(m_x^{(p)}, \Gamma_\nu) = \sum_{(p)} \sum_{(n)} \Phi_{\pi/4}(m_x^{(p,n)}, \Gamma_\nu), \quad (1.3.15)$$

where $m_x^{(p,n)}$ are the components of $m_x^{(p)} \cap \{\bigcup_{\nu=1}^q w^{-1}(\Gamma_\nu(\varepsilon))\}$ containing at least one of the points $z_{i,x}(\Gamma_\nu)$ for $\alpha = \pi/4$.

Two types of segments $m_x^{(p,n)}$ are possible:

(1) The segment $m_x^{(p,n)}$ coincides with $m_x^{(p)}$. i.e. the curve $w(m_x^{(p)})$ lies entirely in $\bigcup_{\nu=1}^q \Gamma_\nu(\varepsilon)$.

(2) $m_x^{(p,n)} \neq m_x^{(p)}$.

In the second case, by the definition of $m_x^{(p,n)}$, its w-image contains an arc joining a point $w(z_{i,x}(\Gamma_\nu)) \in w(m_x^{(p,n)})$ with the boundary of $\bigcup_{\nu=1}^q \Gamma_\nu(\varepsilon)$. Therefore we find $L_{m_x^{(p,n)}} \geq c^*(\varepsilon)$. where $c^*(\varepsilon)$ is the spherical distance between $\bigcup_{\nu=1}^q \Gamma_\nu$ and the boundary of $\bigcup_{\nu=1}^q \overline{\Gamma_\nu(\varepsilon)}$. Further, as for each p there can be at most one arc of form (1).

$$\sum_{(p)} \sum_{(n)} 1 \leq \frac{1}{c^*(\varepsilon)} \sum_{(p)} \sum_{(n)} L_{m_x^{(p,n)}} + \sum_{(p)} 1. \quad (1.3.16)$$

Taking into account (1.3.15) and (1.3.16) and using Lemma 1 for all segments $m_x^{(p,n)}$, by summation we obtain

$$\sum_{(p)} \Phi(D, x, \Gamma_\nu) \leq \frac{8}{\pi} \sum_{\nu} \sum_{(n)} Var_{m_x^{(p,n)}} \alpha_x(y)$$

$$+ h(\Gamma_1, \ldots, \Gamma_q) \sum_{(p)} \sum_{(n)} L_{m_x^{(p,n)}}$$

$$+ \sum_{(p)} \sum_{(n)} 1 + \sum_{(p)} \sum_{(n)} n^*_{m_x^{(p,n)}}$$

$$\leq \frac{8}{\pi} \sum_{(p)} \sum_{(n)} Var_{m_x^{(p,n)}} \alpha_x(y) + \left(h(\Gamma_1, \ldots, \Gamma_q) + \frac{1}{c^*(\varepsilon)} \right)$$

$$\times \sum_{(p)} \sum_{(n)} L_{m_x^{(p,n)}} + \sum_{(p)} 1 + \sum_{(p)} \sum_{(n)} n^*_{m_x^{(p,n)}}$$

$$\leq \frac{8}{\pi} Var_{J_x^{(\varepsilon)}} \alpha_x(y) + h(\Gamma_1, \ldots, \Gamma_q, \varepsilon) \int_{J_x^{(\varepsilon)}} \frac{|w'(z)|}{1 + |w(z)|^2} dy$$

$$+ \sum_{(p)} 1 + \sum_{(p)} \sum_{(n)} n^*_{m_x^{(p,n)}}.$$

Arguing as in Subsection 1.3.3, we derive

$$\sum_{\nu=1}^{q} \int_{x_1}^{x_2} \Phi(D, x, \Gamma_\nu) dx \leq \frac{8}{\pi} \int_{x_1}^{x_2} Var_{J_x^{(\varepsilon)}} \alpha_x(y) dx$$

$$+ h(\Gamma_1, \ldots, \Gamma_q, \varepsilon) \int_{x_1}^{x_2} \int_{J_x^{(\varepsilon)}} \frac{|w'(z)|}{1 + |w(z)|^2} dy dx + \frac{l(D)}{2},$$

and similarly

$$\sum_{\nu=1}^{q} \int_{y_1}^{y_2} \Phi(D, y, \Gamma_\nu) dy \leq \frac{8}{\pi} \int_{y_1}^{y_2} Var_{J_y^{(\varepsilon)}} \alpha_y(x) dy$$

$$+ h(\Gamma_1, \ldots, \Gamma_q, \varepsilon) \int_{y_1}^{y_2} \int_{J_y^{(\varepsilon)}} \frac{|w'(z)|}{1 + |w(z)|^2} dx dy + \frac{l(D)}{2}.$$

The last two inequalities together with (1.2.1) of Section 1.2 complete the proof.

1.4. Estimates of Lengths of Γ-Lines for Angular-Quasiconformal Mappings

1.4.1. Two characteristic properties of analyticity are well known:

(A) A holomorphic function $f(z)$ single-valued in a neighborhood of a point z_0 ($f'(z_0) \neq 0$) carries a "small" disk centered at z_0 into a "small" disk centered at $f(z_0)$.

(B) (Conformity) The angle $\alpha(z_0)$ between two arbitrary lines intersecting at a point z_0 is equal to the angle $\alpha(f(z_0))$ formed by the f-images of these lines at $f(z_0)$.

The classical concept of quasiconformity is a generalization of (A). Its characteristic property is that a quasiconformal mapping carries a "small" disk into a "small" ellipse. Thus, it is natural also to find a generalization of conformity based on (B). We offer the following one: *a function $f(z)$ is "quasiconformal" if there exist some constants $0 < c_1 < c_2$ such that for "almost" every point z*

$$c_1 \leq \frac{\alpha(f(z))}{\alpha(z)} \leq c_2. \tag{1.4.1}$$

Under this condition the more close c_1 and c_2 are to 1, the less $\alpha(f(z))$ differs from $\alpha(z)$, and the case $c_1 = c_2 = 1$ is the usual conformity.

Now consider the class of functions $f(z)$ satisfying the condition: if $\alpha' \leq \alpha(z) \leq \pi/2$ then

$$\alpha \leq \alpha(f(z)) \leq \pi - \alpha, \quad 0 < \alpha < \alpha'. \tag{1.4.2}$$

This condition has an obvious geometric meaning: f-images of sufficiently large angles in z-plane cannot be too small. Note that (1.4.2) does not restrict f-images of angles $\alpha(f(z))$ for small $\alpha(z)$, $\alpha(z) < \alpha'$. Thus, (1.4.2) can be considered as a "generalized quasiconformality" of a function $f(z)$.

Now we turn to exact definitions. According to Stoilov [1, 2] an interior mapping is a mapping that carries any open set into an open set and never a continuum into a single point.

We shall say that a function $w(z)$ defined on $\overline{D}$ is angular-quasiconformal or $w(z) \in M_\alpha(D)$, $0 < \alpha < \pi/4$ if $w(z)$ is an interior mapping such that the following conditions are satisfied in any point $z \in \overline{D}$, with the possible exception of a finite number of singular points from D:

(i) $w(z)$, $z = x + iy$, has continuous second derivatives with respect to x and y;

(ii) if $\pi/4 \leq \alpha(z) \leq \pi/2$, then $\alpha \leq \alpha(w(z)) \leq \pi - \alpha$.

Commenting this definition, notice that interior mappings have the same topological structure as holomorphic functions (see Stoilov [1, 2]). Particularly for interior mappings the pre-images of points are points as well and those of curves are curves. The requirement that $w(z)$ is an interior mapping satisfying (i) is the minimal one permitting to consider the concepts and magnitudes defined in Sections 1.2 and 1.3 without additional difficulties. The restriction on finiteness of (singular) points, where (i) and (ii) can be broken, is similar to the well-known statement that a function meromorphic in $\overline{D}$ does not have more than a finite number of singularities there.

1.4.2. Tangent variation principle for functions of $M_\alpha(D)$. In this subsection we show that the main results of Sections 1.1–1.3 are extendable to the class $M_\alpha(D)$.

Theorem 1. *Let $w(z) \in M_\alpha(D)$ be any function such that $Var_{J_x}\alpha_x(y) < \infty$ for $x_1 < x < x_2$ and $Var_{J_x}\alpha_y(x) < \infty$ for $y_1 < y < y_2$, and let Γ be a smooth Jordan curve with $\nu(\Gamma) < \infty$.*

Then there exists a constant $K(\Gamma, \alpha) < \infty$ depending only on $\nu(\Gamma)$ and α such that

$$L(D, \Gamma) \leq K(\Gamma, \alpha)\left(V(D) + l(D)\right). \tag{1.4.3}$$

Proof. Note that for functions $w \in M_\alpha(D)$ the arguments and relations of Section 1.2 up to inequality (1.2.3) remain valid. By definition, the angle $\widetilde{\alpha}_i(+)$ is the w-image of the angle $\alpha(z) = \pi/2 - \alpha_{i,x}(\Gamma)$ formed by J_x and the tangent to $w^{-1}(\Gamma)$ at $z_{i,x}(\Gamma)$. Since $\alpha_{i,x}(\Gamma) \leq \pi/4$ by the definition, $\pi/4 \leq \alpha(z) \leq \pi/2$. Hence, by $w \in M_\alpha(D)$ it follows that $\alpha \leq \widetilde{\alpha}_{i,x}(+) := \alpha(w(z)) \leq \pi - \alpha$. Similarly, $\alpha \leq \widetilde{\alpha}_{i+1}(-) \leq \pi - \alpha$. The inequality (1.2.13) of Subsection 1.2.3 now can be rewritten in the form

$$Var_{m^{(p)}_{x,i}} \alpha_x(y) \geq \widetilde{\alpha}_i(+) + \widetilde{\alpha}_{i+1}(-) - \nu(\Gamma) \geq 2\alpha - \nu(\Gamma).$$

First suppose that $\nu(\Gamma) \leq \alpha$. Then the last inequality implies

$$Var_{m^{(p)}_{i,x}} \alpha_x(y) \geq \alpha. \tag{1.4.4}$$

Following the scheme by which in Section 1.2 the inequality (1.2.8) was derived from (1.2.2), from (1.4.4) we deduce

$$\Phi_p \leq \frac{1}{\alpha} Var_{m^{(p)}_{i,x}} \alpha_x(y) + 1.$$

Summation of these estimates over all p gives

$$\Phi(D, x, \Gamma) = \sum_{(p)} \Phi_p \leq \frac{1}{\alpha} Var_{J_x} \alpha_x(y) + \sum_{m^{(p)}_x} 1.$$

As in Section 1.2, taking into account the finiteness of singular points of $w(z)$, we obtain

$$\int_{x_1}^{x_2} \Phi(D, x, \Gamma) dx \leq \frac{1}{\alpha} \int_{x_1}^{x_2} Var_{J_x} \alpha_x(y) dx + \frac{l(D)}{2},$$

and similarly

$$\int_{y_1}^{y_2} \Phi(D, y, \Gamma) dy \leq \frac{1}{\alpha} \int_{y_1}^{y_2} Var_{J_y} \alpha_y(x) dy + \frac{l(D)}{2}.$$

Adding these two inequalities and taking into account the inequality (1.2.1) of Section 1.2, we obtain that for $\nu(\Gamma) \leq \alpha$

$$L(D, \Gamma) \leq \frac{\sqrt{2}}{\alpha} V(D) + \sqrt{2} l(D). \tag{1.4.5}$$

If $\nu(\Gamma) < \infty$, then arguing as in Section 1.2 after deriving the inequality
(1.2.15), we choose $\varepsilon_0 = \alpha$ and single out the parts $\Gamma^{(\nu)}$ of the curve Γ. Then
the number of these parts obviously are equal to $[\nu(\Gamma)/\alpha]$ or $[\nu(\Gamma)/\alpha] + 1$.
Summation of inequality (1.4.5) over all ν gives

$$L(D,\Gamma) = \sum_{(\nu)} L(D,\Gamma^{(\nu)}) \leq \sqrt{2}\frac{\nu(\Gamma)+\alpha}{\alpha^2}(V(D)+l(D)),$$

hence denoting $K(\Gamma,\alpha) = \sqrt{2}(\nu(\Gamma)+\alpha)\alpha^{-2}$ and taking into account that
$\alpha \leq \pi/4$, we come to inequality (1.4.3).

1.4.3. Estimates of $\sum_{(\nu)} L(D,\Gamma_\nu)$ for functions of $M_\alpha(D)$. Assuming
that $w \in M_\alpha(D)$ is a given function. in this subsection we presuppose that
the magnitudes $Var_{J_x}\alpha_x(y)$ and $Var_{J_y}\alpha_y(x)$ are bounded for $x_1 < x < x_2$
and $y_1 < y < y_2$ respectively. and also that for the same x and y the integrals

$$L_{J_x} := \int_{J_x} \frac{|w_y'(z)|}{1+|w(z)|^2}dy. \quad L_{J_y} := \int_{J_y} \frac{|w_x'(z)|}{1+|w(z)|^2}dx$$

exist and are bounded.

Theorem 2. *Let $w \in M_\alpha(D)$ be any function satisfying the above conditions
and let $\Gamma_1, \Gamma_2, \ldots, \Gamma_q$ be any disjoint bounded smooth Jordan curves with
$\nu(\Gamma) < \infty$. Then there exists a constant $h = h(\Gamma_1, \Gamma_2, \ldots, \Gamma_q, \alpha)$ such that*

$$\sum_{\nu=1}^{q} L(D,\Gamma_\nu) \leq \frac{2\sqrt{2}}{\alpha}V(D) + h\left(\int_{x_1}^{x_2} L_{J_x}dx + \int_{y_1}^{y_2} L_{J_y}dy\right) + \sqrt{2}l(D).$$

$$(1.4.6)$$

Proof. Observe that Lemma 1 of Subsection 1.3.2 is valid for functions
$w \in M_a(D)$ (even if condition (ii) of the definition of $M_a(D)$ is not satisfied).
We apply Lemma 1 to the curves $\mu := \mu^{(p)} := m_x^{(p)}$. Let $\alpha(z)$ be the smaller
angle between J_x and the tangent to $w^{-1}(\Gamma)$ at $z_{i,x}(\Gamma_\nu)$. Observe that $\alpha(z) =
\pi/2 - \alpha_{i,x}(\Gamma_\nu)$ and $\pi/4 \leq \alpha(z) \leq \pi/2$ as $\alpha_{i,x}(\Gamma_\nu) \leq \pi/4$ by the definition
of points $z_{i,x}(\Gamma_\nu)$. Therefore, $0 < \alpha \leq \alpha(w(z)) \leq \pi - \alpha$ by the definition of
$M_\alpha(D)$. Hence $\Phi_p = \Phi_\alpha\left(m_x^{(p)}, \Gamma_\nu\right)$. and consequently

$$\Phi(D, x, \Gamma_\nu) = \sum_{(p)} \Phi_p = \sum_{(p)} \Phi_\alpha\left(m_x^{(p)}, \Gamma_\nu\right). \qquad (1.4.7)$$

If there are no singular points of w on J_x, then the following equalities are true:

$$\sum_{(p)} \int_0^1 \left| \frac{\partial}{\partial t} \arg \frac{\partial}{\partial t} w \left(\mu^{(p)}(t) \right) \right| dt = \sum_{(p)} Var_{m_x^{(p)}} \alpha_x(y) = Var_{J_x} \alpha_x(y), \quad (1.4.8)$$

$$\sum_{(p)} L_{\mu^{(p)}} = \sum_{(p)} L_{m_x^{(p)}} = L_{J_x}, \quad (1.4.9)$$

and

$$\sum_{(p)} n^*_{m_x^{(p)}} = 0. \quad (1.4.10)$$

From relations (1.4.7) to (1.4.10) and Lemma 1 we find

$$\Phi(D, x, \Gamma_\nu) \le \frac{2}{\alpha} Var_{J_x} \alpha_x(y) + h L_{J_x} + \sum_{m_x^{(p)}} 1. \quad (1.4.11)$$

As the set of singular points of $w(z)$ and the magnitudes in inequality (1.4.11) are finite, integrating from x_1 to x_2 and taking into account the inequality (1.2.5) of Section 1.2 we get

$$\sum_{\nu=1}^q \Phi(D, x, \Gamma_\nu) \le \frac{2}{\alpha} \int_{x_1}^{x_2} Var_{J_x} \alpha_x(y) dx + h \int_{x_1}^{x_2} L_{J_x} dx + \frac{l(D)}{2}.$$

Similarly,

$$\sum_{\nu=1}^q \Phi(D, y, \Gamma_\nu) \le \frac{2}{\alpha} \int_{y_1}^{y_2} Var_{J_y} \alpha_y(x) dy + h \int_{y_1}^{y_2} L_{J_y} dy + \frac{l(D)}{2}.$$

Estimate (1.4.6) follows from (1.2.1) of Section 1.2 and the last two inequalities.

1.4.4. Similarities of the principle of length and mass distribution for functions of $M_\alpha(D)$. In this subsection, we use the notation of Subsections 1.1.3 and 1.1.4.

Theorem 3. *If $w \in M_\alpha(D)$, then*

$$\int_0^\infty \frac{L(D, \Gamma(R))}{\Psi(R)} dR \le \sqrt{2} \int_{x_1}^{x_2} m_\Psi(J_x) dx + \sqrt{2} \int_{y_1}^{y_2} m_\Psi(J_y) dy, \quad (1.4.12)$$

and particularly for $\Psi = 1$

$$\int_0^\infty L(D, \Gamma(R)) dR \le \sqrt{2} \int_{x_1}^{x_2} l(w(J_x)) dx + \sqrt{2} \int_{y_1}^{y_2} l(w(J_y)) dy. \quad (1.4.12')$$

Proof. Note that for functions from $M_\alpha(D)$ the inequalities (1.3.9) and (1.3.9') of Section 1.3 are valid. Consequently,

$$\int_0^\infty \frac{L_x(D, \Gamma(R))}{\Psi(R)} dR = \int_0^\infty \int_{x_1}^{x_2} \sum_{i=1}^{\Phi(D,x,\Gamma(R))} \frac{1}{\Psi(R) \cos \alpha_{i,x}(\Gamma(R))} dx dR$$

$$\le \sqrt{2} \int_{x_1}^{x_2} \int_0^\infty \frac{\Phi(D. x, \Gamma(R))}{\Psi}(R) dR dx, \quad (1.4.13)$$

and

$$\int_0^\infty \frac{L_y(D, \Gamma(R))}{\Psi(R)} dR \le \sqrt{2} \int_{y_1}^{y_2} \int_0^\infty \frac{\Phi(D, y, \Gamma(R))}{\Psi(R)} dR dy. \quad (1.4.13')$$

Arguing as in Subsection 1.1.3, we conclude that

$$\Delta l(w(J_x)) \sim \frac{\Delta R}{\sin \beta_{i,x}(\Gamma(R))}$$

for the length element of the arc $l(w(J_x))$ at the point $Re^{i\theta_{i,x}}$. Since the angle $\beta_{i,x}(\Gamma(R))$ is the w-image of $\pi/2 - \alpha_{i,x}(\Gamma(R))$ and $\pi/4 \le \pi/2 - \alpha_{i,x}(\Gamma(R)) \le \pi/2$, by the definition of $M_\alpha(D)$, we obtain $\alpha \le \beta_{i,x}(\Gamma(R)) \le \pi - \alpha$. Thus, $\Delta l(w(J_x)) \ge \Delta R$ and

$$\Delta m_\Psi(J_x) \ge \frac{\Delta R}{\Psi(R)}.$$

Consequently,

$$\int_{x_1}^{x_2} \int_0^\infty \frac{\Phi(D, x, \Gamma(R))}{\Psi(R)} dR dx \leq \int_{x_1}^{x_2} m_\Psi(J_x \cap D_x) dx \leq \int_{x_1}^{x_2} m_\Psi(J_x) dx$$

and

$$\int_{y_1}^{y_2} \int_0^\infty \frac{\Phi(D, y, \Gamma(R))}{\Psi(R)} dR dy \leq \int_{x_1}^{x_2} m_\Psi(J_y) dy.$$

The inequality (1.4.12) now follows from (1.4.13), (1.4.13′) and (1.1.7) of Subsection 1.1.3.

1.5. Remarks on Application of Estimates of $L(D, \Gamma)$

It was mentioned in the introduction of this book that Γ-lines can be considered as a generalization of the concept of a-points. Also there were mentioned some connections of estimates of Γ-lines with several physical problems, with the principle of length and mass distribution, etc. In this section, we show that investigation of some other problems and various (new or already considered) magnitudes can be reduced to the estimates of $L(D, \Gamma)$.

1.5.1. Estimates of Γ-lines as distortion theorems. Usually the term "distortion theorem" is used for results describing how a geometric object given in the plane z is "distorted" by a mapping f, where f is a function from a given class. The same holds if we substitute the z-plane by the f-plane and the function f by f^{-1}. As some examples, one can mention the classical theorems of Koebe and Ahlfors. Consequently, if for a given curve Γ and a function f from a given class we have an estimate of the length $L(D, \Gamma, f)$ of Γ-lines in terms of the characteristic function of the same class, then we immediately get a typical theorem on the distortion of Γ by the mapping f^{-1}.

Note that, if f is an entire function and Γ is a straight line, one can obtain even an exact two-sided estimate of $L(D(r), \Gamma, f)$ in terms of Ahlfors' characteristic function $A(r, f)$ (see Chapter 4).

1.5.2. Estimates of Γ-lines and distribution of real and imaginary parts of complex functions. Observe that a-points of a function $w(z)$ can be considered as solutions of the following system of equations:

$$\begin{cases} \operatorname{Re} w(z) = A \\ \operatorname{Im} w(z) = B \end{cases} \quad a = A + iB = \text{const} \in \mathbb{C}.$$

On the other hand, the solutions of any of these equations, say $\operatorname{Im} w(z) = B = \text{const}$, can be considered as Γ-lines for $\Gamma = \{w : \operatorname{Im} w = B\}$ or as level sets of the function $\operatorname{Im} w$. Therefore, Γ-lines (or what are the same level sets) in a sense are analogs of a-points of the function $w(z)$, if $\operatorname{Im} w(z)$ is considered instead of $w(z)$. Consequently, the investigation of Γ-lines opens a peculiar way for the construction of the *value distribution theory for imaginary or real parts of complex functions*. At the same time, if it is assumed that $\Gamma = \{w : |w| = A\}$, $A = \text{const} > 0$, then the Γ-lines of a function w can be considered as some analogs of a-points. Therefore, we can similarly study *value distribution theory for functions $|w|$*.

1.5.3. Connection of estimates of $L(D,\Gamma)$ and close-location of a-points. The estimates of $L(D,\Gamma)$ from above contain definite information on the mutual location of a-points for $a \in \Gamma$. Indeed, let the points $a_1, a_2, \ldots$ belong to Γ, and let the function $w(z)$ be angular-quasiconformal in $\overline{D}$. Then the estimates of $L(D,\Gamma)$ indicate a definite smallness of lengths of the arcs l_j, $j = 1, 2, \ldots, \Phi$, which are the components of the set $w^{-1}(\Gamma) = \bigcup_{j=1}^{\Phi} l_j$. Now assume that the total number of $a_1, a_2, \ldots$-points is remarkably larger than the number Φ of arcs l_j. Then, on average, a large number of $a_1, a_2, \ldots$-points lie on the arcs l_j, and since the lengths of these arcs are small on average, one can conclude that $a_1, a_2, \ldots$-points are close to one other.

This scheme is realized as a proximity property of a-points of meromorphic functions in the case of a function $w(z)$ meromorphic in $\mathbb{C}$ or in the disk $D(r)$ (see Chapters 3 and 4). However, the Tangent Variation Principle which establishes the above estimates for $L(D,\Gamma)$ is valid for the remarkably wide classes of functions and domains. Hence, it is possible to test this principle in many known classes of functions, holomorphic, meromorphic, angular-quasiconformal, etc., in order to check the possible close-location of various a-points.

1.5.4. Connections with the Blaschke sums. The Blaschke sum

$$\sum_{z_k \in D(1)} (1 - |z_k|), \tag{1.5.1}$$

is an indicator of the density of zeros z_k (counted according to their multiplicities) of functions holomorphic in the unit disk $D(1)$. In case of an arbitrary domain D, the following generalization of the Blaschke sum seems to be natural:

$$B(D) := \sum_{z_k \in D} \rho(z_k, D), \tag{1.5.2}$$

where $\rho(z_k, D)$ is the distance from z_k to the boundary of D. It is evident that (1.5.2) becomes (1.5.1) if $D = D(1)$. On the other hand, the sum (1.5.2) can be considered also for functions $w(z)$ which are angular-quasiconformal in $\overline{D}$ if as multiplicities of a-points $z_i(a)$ (particularly zeros $z_i(0) = z_i$) we accept the multiplicities of the algebraic or usual points $w(z_i(a))$. Note that the topological structure of interior mappings and meromorphic functions coincides by definition. Therefore, from the fact that the Riemann surface $F_{\overline{D}} = F_{\overline{D}}(w) = \{w(z) : z \in \overline{D}\}$ is closed it follows that its singular points are only the algebraic branching points.

Now we shall show that the estimates of $L(D, \Gamma)$ immediately imply some estimates of the sum (1.5.2). Namely, if $w(z) \neq \infty$ as $z \in D$, then the following estimate is true:

$$B(D) \leq L(D, \gamma^+(0)), \tag{1.5.3}$$

where $\gamma^+(0)$ is the real semi-axis $\{z : \text{Im } z = 0, 0 \leq \text{Re } z \leq \infty\}$. Assume that m_k is the multiplicity of z_k. Then we move from zeros z_k along $\gamma^+(0)$-lines in the directions corresponding to the move from 0 to $+\infty$ on the semi-axis $\gamma^+(0)$. If $F_{\overline{D}}$ has no algebraic branching points lying over $\gamma^+(0)$, then the $\gamma^+(0)$-lines $l_{k,i}$, $i = 1, 2, \ldots, m_k$, outcoming from each point z_k (their number is exactly equal to m_k) reach the boundary of D (if a $\gamma^+(0)$-line would end inside D, then we would have $w(z_0) = \infty$ in its endpoint z_0). Hence

$$\sum_{a_k \in D} \rho(z_k, D) \leq \sum_{(k)} \sum_{i=1}^{m_k} |l_{k,i}| \leq L(D, \gamma^+(0)),$$

where $|l_{k,i}|$ is the length of $l_{k,i}$.

Consider now the case when the Riemann surface $F_{\overline{D}}$ has algebraic branching points lying over $\gamma^+(0)$. We shall show that in this case for each zero z_k of the multiplicity m_k one can single out from the set $w^{-1}(\gamma^+(0))$ exactly the m_k lines outcoming from z_k and ending on the boundary of D. It is clear that the number of $\gamma^+(0)$-lines outcoming from any zero of the multiplicity m_k is exactly m_k. These we consider as the starting points of lines $l_{k,i}$, $i = 1, 2, \ldots, m_k$. While moving on these lines we can meet some multiple points. Let m^* be the multiplicity of such a point $z_\tau(a)$. Then in this point exactly $2m^*$ lines $\gamma^+(0)$ are crossed, m^* of which are w^{-1}-pre-images of the segment $\{w : \operatorname{Im} w = 0, 0 \le \operatorname{Re} z \le a\}$ (we denote them $\widetilde{l}_{\tau,i}(a)$, $j = 1, 2, \ldots, m^*$), and the remaining m^* lines $\gamma^+(0)$ are w^{-1}-pre-images of the semi-axis $\{w : \operatorname{Im} w = 0, a < \operatorname{Re} z \le \infty\}$ (we denote them $\widetilde{\widetilde{l}}_{\tau,i}(a)$, $j = 1, 2, \ldots, m^*$). Arbitrarily connecting a line $\widetilde{l}_{\tau,i}(a)$ with exactly one line $\widetilde{\widetilde{l}}_{\tau,i}(a)$ we determine the continuation of lines through multiple points. Continuing this process, in accordance with the emerging new multiple points along the lines $l_{k,i}$, for each zero of multiplicity m_k, we single out exactly the m_k lines $l_{k,i}$, $i = 1, 2, \ldots, m_k$, outcoming from z_k and ending on the boundary of D. Thus, inequality (1.5.3) is valid also when there exist algebraic branching points lying over $\gamma^+(0)$.

1.5.5. Gelfond's magnitudes $\Phi(r, \Gamma)$.

They are the numbers of the points z_i for which $w(z_i) \in \Gamma$ on the circle $|z| = r$. A. O. Gelfond introduced these magnitudes in 1934 and obtained some estimates of $\Phi(r, \Gamma)$ for the case when w is an entire function and Γ is a straight line. Later, a series of works appeared, where some estimates of $\Phi(r, \Gamma)$ were established under the assumption that Γ is a circle or an algebraic curve (see Chapter 4).

For wide classes of curves Γ, the magnitudes $\Phi(r, \Gamma)$ can be investigated by the use of the Tangent Variation Principle, since the following simple connection between Φ and L is true:

$$\int_{r_0}^r \Phi(t, \Gamma) dt \le L(r, \Gamma) - L(r_0, \Gamma). \tag{1.5.4}$$

Indeed, the set of points $r \in R^+$, each corresponding to the point $z = re^{i\varphi} \in w^{-1}(\Gamma) \cap \{z : r_0 \le |z| \le r\}$, compose a collection of segments and semi-intervals which are in one-to-one correspondence to the arcs of set $w^{-1}(\Gamma) \cap \{z : r_0 \le |z| \le r\}$. Let $\overline{L}(r, r_0, \Gamma)$ be the total length of these segments and semi-intervals (i.e. the length of the projection of this set on the real semi-axis with respect to the multiplicity of projection). Obviously

$$\int_{r_0}^r \Phi(t, \Gamma) dt \le \overline{L}(r, r_0, \Gamma) \tag{1.5.5}$$

and

$$\overline{L}(r, r_0, \Gamma) \le L(r, \Gamma) - L(r_0, \Gamma), \tag{1.5.6}$$

whence (1.5.4) holds.

Note that the magnitude $\Phi(r,\Gamma)$ is an upper bound for the number of those lines which are asymptotic and have common points with the disk $D(r)$.

1.5.6.   One more connection with asymptotic behaviors of functions. Let $J(r,\Gamma,k)$, $0 < k < 1$, be the number of those connected components of the set $w^{-1}(\Gamma) \cap D(r)$, which have common points with both boundary circles of the ring $\{z : kr \leq |z| \leq r\}$. This magnitude is an upper bound for the number of asymptotic Γ-lines intersecting with the same circles.

Investigation of magnitudes $J(r,\Gamma,k)$ is reducible to the estimates of $L(r,\Gamma)$ also: since the length of each mentioned component is not less than $(1-k)r$. Hence

$$J(r,\Gamma,k) \leq \frac{L(r,\Gamma) - L(kr,\Gamma)}{(1-k)r}. \tag{1.5.7}$$

Obviously, instead of the rings one can consider other doubly-connected domains for which some similarities of (1.5.7) are valid.

1.5.7.   The connection between estimates of lengths $L(D,\Gamma,w)$ and estimates of derivatives of $w(z)$ is very simple. Obviously, the above estimates of the lengths $L(D,\Gamma,w)$ (Tangent Variation Principle) lead to estimates from below of the magnitudes $|w'(z)|$ on Γ-lines. Application of this remark to functions $w(z)$ meromorphic in $\mathbb{C}$ leads to the exact estimates from below for the magnitude $|w'(z)|$ on the sets of simple a-points of the function w (see Section 2.3; also the author's paper [14]).

CHAPTER 2

NEVANLINNA AND AHLFORS' THEORIES: ADDITIONS

2.1. Basic Concepts and, Outcomes of Nevanlinna Value Distribution Theory and Ahlfors' Theory of Covering Surfaces

2.1.1. Nevanlinna's deficiency relation. In the remainder of this chapter, we denote by M_R the set of non-constant functions meromorphic in $|z| < R \leq \infty$ and by M the set M_∞ of functions meromorphic in the complex plane $\mathbb{C}$.

Let $P_n(z)$ be a polynomial of degree n. According to Main Theorem of Algebra, the number of solutions (multiplicities are accounted) of the equation $P_n(z) = a \in \mathbb{C}$ is equal to n. If we denote by $n(r, a) := n(r, a, f)$ the number of a-points of a function f meromorphic in the disk $|z| \leq r$, then the assertion of Main Theorem of Algebra can be written in the form

$$\frac{n(r, a, P_n)}{n} = \frac{n(r, b, P_n)}{n} = 1, \ r > r_0, \tag{2.1.1}$$

where $a, b \in \mathbb{C}$ are arbitrary and the number r_0 depends on a, b and P_n.

The main results of the Value Distribution Theory in a certain extent answer the question: *in what form relation (2.1.1) can be extended to functions meromorphic in $\mathbb{C}$?* Note that one cannot expect the relation (2.1.1) to be true for such functions since, for instance, for e^z, we have $n(r, 0, e^z) \equiv 0$, $r \in R^+$. Obviously, there must be a characteristic of $w(z) \in M_\infty$ that plays the same role as degree n of a polynomial. Also note that such a characteristic cannot be a constant since in general $n(r, a, w) \to \infty$ as $r \to \infty$. Such a magnitude in the theory of meromorphic functions is Nevanlinna's characteristic function

$$T(r) := T(r, w) := m(r, \infty) + N(r, \infty),$$

where

$$m(r, \infty) := m(r, \infty, w) := \frac{1}{2\pi} \int_0^{2\pi} \ln^+ |w(re^{i\varphi})| d\varphi$$

(henceforth, we assume that $\ln^+ x$ is equal to $\ln x$ for $x \geq 1$ and vanishes for $0 \leq x < 1$), and

$$N(r, a) := N(r, a, w) := \int_0^r \frac{n(t, a, w) - n(0, a, w)}{t} dt + n(0, a, w) \ln r, \quad a \in \overline{\mathbb{C}}.$$

Nevanlinna's classical Deficiency Relation for meromorphic functions establishes the following analog of (2.1.1): *there exists a no more than countable set $P \subset \overline{\mathbb{C}}$, such that for any points $a, b \notin P$*

$$\limsup_{r \to \infty} \frac{N(r, a, w)}{T(r)} = \limsup_{r \to \infty} \frac{N(r, b, w)}{T(r)} = 1. \tag{2.1.2}$$

The observed difference between relations (2.1.1) and (2.1.2) follows from the nature of meromorphic functions. The magnitude $\lim_{r \to \infty} \sup N(r, a)/T(r)$ can be strictly less than 1 for some $a = a^* \in P$, and such a^* are not typical; they are "deficient" for functions $w(z) \in M$. This is the nature of deficient values. They are defined as:

$$\delta(a) := 1 - \limsup_{r \to \infty} \frac{N(r, a)}{T(r)}$$

and are called "the deficiencies" of $w(z)$ in $a \in \overline{\mathbb{C}}$. The value a is called "deficient value", if $\delta(a) > 0$ (i.e. $\limsup_{r \to \infty} N(r, a)/T(r) < 1$) and "not deficient value", if $\delta(a) = 0$ (i.e. $\limsup_{r \to \infty} N(r, a)/T(r) = 1$).

According to the deficiency relation, for any $a \notin P \subset \overline{\mathbb{C}}$, where P is not more than a countable set,

$$\delta(a) = 0 \tag{2.1.3}$$

(this coincides with (2.1.2)) and

$$\sum_{(a)} \delta(a) \le 2. \tag{2.1.4}$$

Thus, if $\delta(a) > 0$, then the magnitude of the deficiency $\delta(a)$ indicates a certain measure of the deviation of the function $N(r, a)$ from the average norm (equal to $T(r)$), and the inequality (2.1.4) shows the upper bound of the sum of all such deviations for all $a \in \overline{\mathbb{C}}$. So, the relation (2.1.3) establishes a remarkable similarity in the distribution of the numbers of the a-points for the majority of values $a \in \overline{\mathbb{C}}$.

2.1.2. The Deficiency Relation is the main consequence of Nevanlinna's First and Second Fundamental Theorems. Below, we omit the proofs of these theorems as they are given in detail in numerous well-known monographs.[2]

---

[2]See Nevanlinna [1], Tsuji [1], Hayman [2], Goldberg and Ostrovskii [1], Petrenko [1], and also Stoilov [1], Markushevich [1], Evgrafov [1], Dinghas [1,2].

First Fundamental Theorem of Nevanlinna. *Let $w(z) \in M_R$, $w(z) \not\equiv$ const. Then for any $a \in \overline{\mathbb{C}}$*

$$m(r,a) + N(r,a) = T(r) + O(1), \quad r \to R, \tag{2.1.5}$$

where

$$m(r,a) := m(r,a,w) := \frac{1}{2\pi} \int_0^{2\pi} \ln^+ \frac{1}{|w(re^{i\varphi}) - a|} d\varphi$$

is Nevanlinna's proximity function.

Note that by (2.1.5)

$$\delta(a) := 1 - \lim_{r \to \infty} \sup \frac{N(r,a)}{T(r)} = \lim_{r \to \infty} \inf \frac{m(r,a)}{T(r)}.$$

On the other hand, relation (2.1.5) states a remarkable invariance property of the expression $m(r,a) + N(r,a)$ with respect to a, within a bounded magnitude.

It appears that each of the magnitudes $m(r,a)$ and $N(r,a)$ can be close to $T(r)$ for given a and $r < R$. The situation changes if we consider collections of pairwise different complex numbers $a_1, a_2, \ldots, a_q$ for $w(z) \in M$. Then it turns out that for the majority of ν the magnitude $N(r,a)$ is close to $T(r)$ and the magnitude of $m(r,a_\nu)$ is negligibly small in comparison to $T(r)$. This conclusion is a consequence of

Second Fundamental Theorem of Nevanlinna. *For $q \geq 3$*

$$\sum_{\nu=1}^{q} m(r,a_\nu) + N_1(r) \leq 2T(r) + O(T(r)), \quad as \quad r \to \infty, \quad r \notin E, \tag{2.1.6}$$

where

$$N_1(r) = N(r,0,w') + (2N(r,\infty,w) - N(r,\infty,w'))$$

and $E \subset [0,\infty)$ is a set of finite Lebesgue measure.

The magnitude $N_1(r)$ has a very simple meaning since $n_1(r) := n(r,0,w') + (2n(r,\infty,w) - n(r,\infty,w'))$ is the sum of orders of all multiple points of the function $w(z)$, which lie in the disk $|z| \leq r$, i.e. each point of the multiplicity k is counted $k-1$ times. By the First Fundamental Theorem, the inequality (2.1.6) can be rewritten in the following forms:

$$\sum_{\nu=1}^{q} (T(r) - N(r,a_\nu)) + N_1(r) \leq 2T(r) + O(T(r)), \quad r \to \infty, \quad r \notin E,$$

$$\tag{2.1.7}$$

$$\sum_{\nu=1}^{q} N(r, a_\nu) - N_1(r) \geq (q-2)T(r) + O(T(r)), \quad r \to \infty, \quad r \notin E. \quad (2.1.7')$$

Further, we shall establish a series of relations similar to (2.1.6) and (2.1.7), from which appropriate "deficiency relations" will follow. We shall omit the proofs of the new deficiency relations as their proof is similar to the below given Nevanlinna's proof of deficiency relation based on (2.1.6) and (2.1.7). Dividing (2.1.6) and (2.1.7) by $T(r)$ and passing to the lower limit as $r \to \infty$, one can obtain

$$\sum_{\nu=1}^{q} \delta(a_\nu) + \Theta_1 \leq 2, \quad (2.1.8)$$

where $\Theta_1 := \lim_{r \to \infty} \inf N_1(r)/T(r)$. Since q is arbitrary, it follows from (2.1.8) that the number of deficient values a with $\delta(a) \in (2^{-(n+1)}, 2^{-n}]$ is not greater than 2^{n+2}. Therefore, the set of deficient values (where $\delta(a) > 0$) is not more than countable (since this set can be represented as a countable sum of finite sets). Consequently, it is possible to put in (2.1.8) ∞ instead of q, so that (2.1.8) becomes (2.1.4).

2.1.3. The observed closeness between $N(r, a)$ and $T(r)$ for the majority of values $a \in \mathbb{C}$ can also be obtained from the following useful identity of Cartan: if $w(z) \in M_R$ and $w(z) \neq$ const, then there exists a constant C independent on r, such that

$$T(r) = \frac{1}{2\pi} \int_0^{2\pi} N(r, e^{i\vartheta}) d\vartheta + C. \quad (2.1.9)$$

If $w \in M$ is a transcendental function, then evidently

$$\lim_{r \to \infty} \frac{T(r)}{\ln r} = +\infty. \quad (2.1.10)$$

Hence the magnitude of the constant C is not essential in (2.1.9). This leads to the qualitative conclusion about the closeness between $N(r, e^{i\vartheta})$ and $T(r)$ in average.

2.1.4. There is an exact statement in Nevanlinna's theory (see Nevanlinna [1], Section 2.2.6) related to the behavior of Valiron's deficient values:

$$\tilde{\delta}(a) := \lim_{r \to \infty} \sup \frac{m(r, a)}{T(r)} = 1 - \lim_{r \to \infty} \inf \frac{N(r, a)}{T(r)}.$$

For any $a \in \overline{\mathbb{C}}$, except at most a set of zero capacity,

$$m(r, a) = O\left[(T(r))^{(1+\varepsilon)/2}\right] \quad \text{as} \quad r \to \infty, \tag{2.1.11}$$

where $0 < \varepsilon = \text{const} < 1/2$. Consequently, for the same values of a

$$\widetilde{\delta}(a) = 0. \tag{2.1.12}$$

2.1.5. The magnitude

$$A(r) := \frac{1}{\pi} \int\int_{|z|<r} \frac{|w'(z)|^2}{(1+|w(z)|^2)^2} r \, dr \, d\varphi. \quad z = re^{i\varphi}, \tag{2.1.13}$$

called Shimizu–Ahlfors characteristic function, has an interesting interpretation. One comes to this interpretation by projecting the w-image of the disk $|z| < r$ (which is same as the Riemann surface $F_r = \{w(z) : |z| \le r\}$) on the Riemann sphere. On this sphere the element of spherical length

$$\frac{|dw|}{(1+|w|^2)} = \frac{|w'(z)|}{(1+|w(z)|^2)} ds$$

and the spherical area element

$$\frac{|w'(z)|^2}{(1+|w(z)|^2)^2} r \, dr \, d\varphi$$

correspond to the usual length element $|w'(z)|ds$ and to the usual area element $|w'(z)|^2 r \, dr \, d\varphi$ respectively. Hence $A(r)$ represents the spherical area of the surface F_r, divided by π (which is the same as the area of the spherical image F_r^* of the surface F_r, divided by π). If F_r^* n times completely covers the Riemann sphere, then $A(r)$ is exactly equal to n, since

$$A(r) = \frac{\text{area of } F_r^*}{\pi} = \frac{n(\text{area of Riemann sphere})}{\pi} = n$$

as the area of the Riemann sphere is equal to π. Thus, in this case the magnitude $A(r) = n$ indicates the number of covering sheets or the multiplicity of coverings of the sphere. In general case, this argument allows to consider the magnitude $A(r)$ as an average number of coverings of the Riemann sphere by the surface F_r^*.

The following useful relation is valid:

$$T^0(r) := \int_0^r \frac{A(t)}{t} dt = T(r) + C, \qquad (2.1.14)$$

where $|C| <$ const. The left-hand side of this relation describes the asymptotic behavior of meromorphic functions in terms of the structure of the Riemann surface F_r^* which covers the Riemann sphere $A(r)$ times on average. The right-hand side of (2.1.14) does the same in terms of the fundamental magnitude $T(r)$. Thus, (2.1.14) indicates a remarkable connection between magnitudes of different nature.

Ahlfors [2] and Shimizu [1] established a similarity of Nevanlinna's first fundamental theorem. In terms of $T^0(r)$, their statement takes the form of following elegant identity:

$$m^0(r, a) + N(r, a) = T^0(r), \qquad (2.1.15)$$

where

$$m^0(r, a) = \frac{1}{2\pi} \int_0^{2\pi} \ln \frac{1}{k(w(re^{i\varphi}), a)} d\varphi - \ln \frac{1}{k(w(0), a)},$$

and $k(w_1, w_2)$ is the chordal distance of points w_1 and w_2 of the Riemann sphere.

2.1.6. In this section, we state some results often used in Nevanlinna's theory. The proofs are omitted since one can find them in numerous monographs and textbooks.

For proving his Second Fundamental Theorem, Nevanlinna established the following relation for any function $w \in M_R$: if a_i, $i = 1, 2, \ldots, q$, are arbitrary pairwise different points, then

$$\sum_{\nu=1}^q m(r, a_\nu, w) \leq m(r, 0, w') + O\left(1 + \sum_{\nu=1}^q m\left(r, \frac{w'}{w - a_\nu}\right)\right), \quad r \to R.$$
$$(2.1.16)$$

By this the proof of the Second Fundamental Theorem was reduced to proving some estimates for the magnitudes $m(r, w'/w - a)$, since $m(r, 0, w') \leq 2T(r) + O(1 + m(r, w'/w))$ as $r \to R$. The mentioned estimates are stated as the well-known Lemma on Logarithmic Derivative, which we give in a form suitable for further applications: Let $w(z) \in M$ be an arbitrary function. Then for any r and R $(r < R)$ and any point $a \in \mathbb{C}$

$$m\left(r, \frac{w'}{w - a}\right) \le 2\ln^+ T(R, w) + 2\left(\ln^+ \frac{1}{R - r} + \ln^+ R\right) + O(1),$$
$$r \to \infty. \tag{2.1.17}$$

As a corollary, one can state that if the function $w(z)$ is of finite order,[3] then

$$m\left(r, \frac{w'}{w - a}\right) = O(\ln r) \quad \text{as} \quad r \to \infty, \tag{2.1.18}$$

and if $w(z)$ is of infinite order, then

$$m\left(r, \frac{w'}{w - a}\right) = O(\ln T(r, w) + \ln r) \quad \text{as} \quad r \to \infty, \quad r \notin E_0, \tag{2.1.19}$$

where E_0 is a set of finite measure in $(0, \infty)$. Thus, for $r \notin E_0$ the sums in the right-hand side of (2.1.16) are negligibly small in comparison with $T(r, w)$, and the magnitude $m(r, 0, w')$ takes a "uniting role" as it estimates the left-hand side sums for arbitrary collections of points $a_1, a_2, \ldots, a_q$.

The following Nevanlinna, Poisson–Jensen and Jensen formulas are frequently used for the investigation of meromorphic functions. Let $f(z) \not\equiv 0$ be a function meromorphic in a disk $|z| \le R$, and let a_m and b_n be its zeros and poles, respectively. Nevanlinna's formulas are:

$$\ln f(z) = \frac{1}{2\pi} \int_0^{2\pi} \ln |f(Re^{i\vartheta})| \frac{Re^{i\vartheta} + z}{Re^{i\vartheta} - z} d\vartheta$$
$$- \sum_{|a_m| < R} \ln \frac{R^2 - \bar{a}_m z}{R(z - a_m)} + \sum_{|b_n| < R} \ln \frac{R^2 - \bar{b}_n z}{R(z - b_n)} + iC, \tag{2.1.20}$$

---

[3]If $\Psi(x) > 0$ is a real-valued function on $(0, \infty)$, then its order ρ_Ψ and lower order λ_Ψ are defined to be the magnitudes

$$\lim_{r \to \infty} \sup \frac{\ln \Psi(x)}{\ln x} \quad \text{and} \quad \lim_{r \to \infty} \inf \frac{\ln \Psi(x)}{\ln x}$$

respectively. By definition, the order ρ and the lower order λ of a meromorphic function are those of $\Psi(r) = T(r)$, where $T(r)$ is Nevanlinna's characteristic function. From the definitions of characteristic functions $A(r)$ and $T(r)$ and from (2.1.14) it follows that the orders and lower orders of $A(r)$ and $T(r)$ coincide, i.e. $\rho_T = \rho_A$ and $\lambda_T = \lambda_A$.

where C is a real constant and

$$\frac{d}{dz}\ln f(z) = \frac{1}{2\pi}\int_0^{2\pi}\ln|f(Re^{i\vartheta})|\frac{2Re^{i\vartheta}}{(Re^{i\vartheta}-z)^2}d\vartheta$$

$$+ \sum_{|a_m|<R}\left(\frac{\bar{a}_m}{R^2-\bar{a}_m z}+\frac{1}{z-a_m}\right)$$

$$- \sum_{|b_n|<R}\left(\frac{\bar{b}_n}{R^2-\bar{b}_n z}+\frac{1}{z-b_n}\right). \qquad (2.1.21)$$

Poisson–Jensen's formula is

$$\ln|f(z)| = \frac{1}{2\pi}\int_0^{2\pi}\ln|f(Re^{i\vartheta})|\mathrm{Re}\frac{Re^{i\vartheta}+z}{Re^{i\vartheta}-z}d\vartheta$$

$$- \sum_{|a_m|<R}\ln\left|\frac{R^2-\bar{a}_m z}{R(z-a_m)}\right|+\sum_{|b_n|<R}\ln\left|\frac{R^2-\bar{b}_n z}{R(z-b_n)}\right|. \quad (2.1.22)$$

Jensen's formula is

$$\ln|c_\lambda| = \frac{1}{2\pi}\int_0^{2\pi}\ln|f(Re^{i\vartheta})|d\vartheta$$

$$- \sum_{0<|a_m|<R}\ln\frac{R}{|a_m|}+\sum_{0<|b_n|<R}\ln\frac{R}{|b_n|}-\lambda\ln R, \quad (2.1.23)$$

where λ is the entire number from the Laurent's expansion $f(z) = c_\lambda z^\lambda + c_{\lambda+1}z^{\lambda+1}+\cdots$, $c_\lambda \neq 0$ in the neighborhood of $z=0$.

2.1.7. Probably, the simplest description of the nature of Ahlfors' Theory of Covering Surfaces can be given by the following interpretation. Observe that if $w(z)$ is meromorphic in the disk $D(r) := \{z:|z|\leq r\}$ and has a given number of a-points in $D(r)$, then there is exactly the same number of points on the Riemann surface $F_r = \{w(z):|z|\leq r\}$ that are projected to a. And if an a-point $z_i(a)\in D(r)$ has the multiplicity $k>1$, then the corresponding $w(z_i(a))\in F_r$ is an algebraic branching point of the same multiplicity k. If $k=1$, then we have a simple a-point $z_i(a)$ and a simple point $w(z_i(a))$. Obviously, there exists a small enough neighborhood σ_a of a, such that the σ_a-neighborhood of each algebraic branching point $w(z_i(a))$ of the multiplicity k completely covers the σ_a-neighborhood of a exactly k

times, and for $k = 1$ the σ_a-neighborhood of the point $w(z_i(a))$ (which is now simple) covers the σ_a-neighborhood of a only once. Hence, the number $n(r, a)$ of a-points in $D(r)$ is equal to the number of coverings of the point a by the Riemann surface F_r or, which is the same, it is equal to the number of complete coverings of "small" σ_a-neighborhoods of a by F_r. As to the number $n_1(r, a)$, it is a "branching measure" of the surface F_r over the point a or over its σ_a-neighborhood.

Thus, it is possible to treat Nevanlinna's Second Fundamental Theorem as a statement about coverings of given points $a_1, a_2, \ldots, a_q$ or their small neighborhoods by the Riemann surface F_r.

In his already classical theory of Covering Surfaces, Ahlfors[4] introduced some concepts which were used for a deep investigation of the nature of the discussed coverings and, particularly, for the investigation of coverings of a given domain D (which is more than the coverings of a-points and their small neighborhoods). What is remarkable is that his theory permits to obtain several generalizations of the main conclusions of Nevanlinna's theory by metrical–topological methods. This leads to a perfectly clear separation of topological and metrical aspects of the Value Distribution Theory. On the other hand, a doubtless advantage of Ahlfors' theory is that it studies the nature of coverings only in terms of covering surfaces (Riemann surfaces), without exploiting functions that generate these surfaces. This allows to extend the main conclusions of the Value Distribution Theory to substantially wider classes of mappings.

2.1.8. We formulate the basic concepts of Ahlfors' theory in particular cases suitable for further applications. We consider only the case when Ahlfors' basic surface is the Riemann sphere.

A triangle on the Riemann sphere is defined as a simply connected domain bounded by three Jordan arcs (sides), which in the remainder of this chapter will be supposed to be smooth. A covering surface F over the Riemann sphere is a surface consisting of triangles glued along their common sides, such that any two neighboring triangles of F are projected to neighboring triangles on the Riemann sphere. A covering surface F is called finite if it consists of finite number of triangles. The last concept is useful particularly since the surface F_r^* (the spheric image of F_r) considered in Section 4.1.6 is actually a finite covering surface over the Riemann sphere.

---

[4]see Ahlfors [3], also Nevanlinna [1], Tsuji [1], Stoilov [1], Hayman [2].

Let D be a domain on the Riemann sphere and let β be a curve on the same sphere. Further, let $J_0(D)$ be the spherical area of D, let $L_0(\beta)$ be the spherical length of β and J be the area of the surface F. The magnitude $s = J/\pi$ is called "average number of sheets of the surface F" (compare with Section 4.1.5). If $J(D)$ is the sum of areas of all parts of F lying over D, then the magnitude

$$s(D) := \frac{J(D)}{J_0(D)}$$

is called "number of sheets of the surface F lying over D". The average number of sheets $s(\beta)$ of F lying over a given curve β on the Riemann sphere is defined similarly: if $L(\beta)$ is the sum of lengths of all arcs from F lying over β, then

$$s(\beta) := \frac{L(\beta)}{L_0(\beta)}.$$

A more detailed consideration of the structure of covering surfaces over a domain D leads to a deep analogy between the magnitudes $s(D)$ and $s(\beta)$ and the sums $m(r, a) + N(r, a)$ of Nevanlinna theory. A connected part of F that lies over D is called an "island" if the projection of the boundary of this part on D does not intersect with D, otherwise it is called "peninsula". The *multiplicity* of an island is the number of its sheets, and *the order* of an island is less than its multiplicity by 1. Further, $n(D)$ is the sum of multiplicities of all islands over D, $n_1(D)$ is the sum of their orders and $m(D)$ is the sum of areas of all peninsulas (where multiplicities of covering is counted) divided by the area of D.

A connected part of the surface F, which lies over β and does not contain points of the boundary ∂F, is a natural similarity of island, and we call it "island over β". Otherwise, such a part of F is a similarity of peninsula, and we call it "peninsula over β". We denote by $n(\beta)$ the sum of multiplicities of all islands over β, by $m(\Gamma)$ the sum of lengths of all peninsulas (in accordance with the multiplicity of covering) divided by the length of β, and by $n_1(\beta)$ the sum of orders of all islands.

Obviously $s(D) := n(D) + m(D)$ and $s(\beta) = n(\beta) + m(\beta)$.

2.1.9. Main results on finite covering surfaces

Theorem A. *Let F be a finite covering surface. Further, let D be a simply connected domain and β be a smooth Jordan curve, both on the Riemann* sphere. If L is the length of the boundary of F in the spherical metric, then

$$S(D) := n(D) + m(D) = S + hL, \qquad (2.1.24)$$

where $|h| < h(D) = \text{const}$, and

$$S(\beta) := n(\beta) + m(\beta) = S + hL, \qquad (2.1.25)$$

where $|h| < h(\beta) = \text{const}$.[5]

Theorem B. *Let F be a finite covering surface, let D_ν, $\nu = 1, 2, \ldots, q$, be a finite collection of simply connected domains with no overlapping closures and let β_ν, $\nu = 1, 2, \ldots, q$, be a finite collection of non-interesting smooth Jordan curves on the Riemann sphere. Then*

$$\sum_{\nu=1}^{q} n(D_\nu) - \sum_{\nu=1}^{q} n_1(D_\nu) \geq (q-2)S - h(D_1, \ldots, D_q)L \qquad (2.1.26)$$

and

$$\sum_{\nu=1}^{q} n(\beta_\nu) - \sum_{\nu=1}^{q} n_1(\beta_\nu) \geq (q-2)S - h(\beta_1, \ldots, \beta_q)L, \qquad (2.1.27)$$

where $h = h(D_1, \ldots, D_q) = \text{const} < \infty$ and $h = h(\beta_1, \ldots, \beta_q) = \text{const} < \infty$.

Ahlfors used somewhat different forms of inequalities (2.1.26) and (2.1.27). Nonetheless, (2.1.26) and (2.1.27) are simple consequences of his first and second fundamental theorems.

Let $n_0(D)$ be the number of simple islands of the surface F lying over the domain D, and let $\overline{n}(D)$ be the number of its islands, where the multiplicities are not counted. Since $n(D) := \overline{n}(D) + n_1(D)$, from (2.1.24) and (2.1.26) we come to the estimate

$$\sum_{\nu=1}^{q} \overline{n}(D_\nu) \geq (q-2)S - h(D_1, \ldots, D_q)L.$$

Evidently $\overline{n}(D) \leq n_0(D) + n_1(D)$. Therefore, once more using (2.1.24) and (2.1.26) we come to the main conclusion of Ahlfors' theory related to the number of simple islands:

$$\sum_{\nu=1}^{q} n_0(D_\nu) \geq (q-4)S - h(D_1, \ldots, D_q)L. \qquad (2.1.28)$$

---

[5]We gave a somewhat rough version of Ahlfors' results since he considered "length of relative boundary" rather than L.

The similarity between the relations (2.1.24) and (2.1.28) and the main results of Nevanlinna theory is evident and it becomes more perfect if the magnitude L is essentially smaller than S.

2.1.10. The mentioned smallness of L with respect to S arises in a series of important cases which we describe below. Let $w(z) \in M_R$. Instead of a surface F we consider the spherical image F_r^*, $r < R$. Then obviously $S := A(r)$. Further, we write $L(r)$, $n(r, D)$, $m(r, D)$, $n_1(r, D)$, $n(r, \beta)$, $m(r, \beta)$, $n_1(r, \beta)$, $n_0(r, D)$ instead of $L, n(D), m(D), n_1(D), n(\beta), m(\beta), n_1(\beta), n_0(D)$.

The following remarkable statement is true.

Lemma of Ahlfors. *If $R = \infty$, then for any ε, $0 < \varepsilon < 1/2$, the inequality*

$$L(r) < A^{1/2+\varepsilon}(r) \tag{2.1.29}$$

is valid outside of a set E of finite logarithmic length, such that

$$\int_E d\ln r < \infty.$$

If $R < \infty$ and

$$\limsup_{r \to \infty} (R - r) A(r) = \infty, \tag{2.1.30}$$

then there exists a sequence $r_n \to R$ such that

$$\frac{L(r_n)}{A(r_n)} \to 0. \tag{2.1.31}$$

Hence, if $F = F_r^*$, from (2.1.24) to (2.1.28) we obtain the relations

$$n(r, D) + m(r, D) = A(r) + hL(r) = A(r) + o(A(r)), \tag{2.1.32}$$

$$n(r, \beta) + m(r, \beta) = A(r) + hL(r) = A(r) + o(A(r)), \tag{2.1.33}$$

$$\sum_{\nu=1}^{q} n(r, D_\nu) - \sum_{\nu=1}^{q} n_1(r, D_\nu) \geq (q - 2)A(r) + hL(r)$$
$$= (q - 2)A(r) + o(A(r)), \tag{2.1.34}$$

$$\sum_{\nu=1}^{q} n(r,\Gamma_\nu) - \sum_{\nu=1}^{q} n_1(r,\Gamma_\nu) \;\geq\; (q-2)A(r) + hL(r)$$

$$= (q-2)A(r) + o(A(r)), \qquad (2.1.34')$$

$$\sum_{\nu=1}^{q} n_0(r,D_\nu) \geq (q-4)A(r) + hL(r) = (q-4)A(r) + o(A(r)), \quad (2.1.35)$$

which are true for $r \to \infty$, $r \notin E$, in the case $R = \infty$, and for $r = r_n \to R$ under the additional condition (2.1.30) in the case when $R < \infty$.

Remark. Lemma of Ahlfors and the relations (2.1.32)–(2.1.35) are valid also for K-quasiconformal mappings defined in $|z| < R \leq \infty$.

This problem is the initial point for discovering some new geometrical properties of the behavior of meromorphic functions. These properties are given in the next section.

2.2. Geometric Deficient Values

2.2.1. The following is the main qualitative corollary of Nevanlinna's First Fundamental Theorem: if a function $w(z)$ meromorphic in $\mathbb{C}$ takes a value $a \in \mathbb{C}$ rarely in the disks $|z| \leq r$, so that a is deficient for $w(z)$, then there are some regions on the circles $|z| = r$, where $|w(z) - a|$ is "small", i.e. for the deficient values a we observe a certain closeness between $w(z)$ and a. It appears that deficiency of a leads to another geometric behavior of the curve $w(z)$, where z belongs to $|z| = r$: the curve is strongly revolved, "coiled" around the point a.

In this section, we establish an analog of the First Fundamental Theorem, in which the role of $m(r,a)$, characterizing the mentioned closeness between $w(z)$ and a, takes a function $\nu(r,a)$ characterizing the extent of "the coilings" around a.

In Section 2.1.11 we already mentioned the absence of an analog of the function $m(r,a)$ in Ahlfors' theory and correspondingly the absence of an analog of the First Fundamental Theorem for values a. We gave also Ahlfors' argument explaining why it is impossible to derive such an analog from the First Fundamental Theorem for domains $D \ni a$. The function $\nu(r,a)$ and the further established analog of the First Fundamental Theorem fill this gap in

Ahlfors' theory. New "geometrical deficient values" will be defined in terms of $\nu(r,a)$ and the corresponding deficiency relation will be established. In its turn, this permits to give a novel interpretation to all results related to deficient values of meromorphic functions in terms of "coiling" around a.

2.2.2. Definitions of $\nu(r,a)$ and $\nu(F,a)$ and main results. Consider the part of the boundary $\partial F_r = \{w(z) : |z| = r\}$, which is over the disk $|w - a| < 1$ for $a \neq \infty$ and over $|w| > 1$ for $a = \infty$. This set is a union of a collection of curves γ_a. We denote by $2\pi\nu_{\gamma_a}$ the increment of $\arg\left(1/\left(w(z) - a\right)\right)$ on γ_a, in the case when $a \neq \infty$, and the increment of $\arg w(z)$ in the case when $a = \infty$. Further, we set

$$\nu(r,a) = \nu(r,a,w) = \sum_{(\gamma_a)} [\nu_{\gamma_a}]',$$

where $[x]'$ is the entire part of x and the sum is taken over all γ_a.

The magnitude of $\nu(r,a)$ is provided by the structure of ∂F_r in the neighborhood of the point a. On the other hand, $\nu(r,a)$ characterizes this structure, as it indicates the total number of complete turns of arcs γ_a around the point a.

A similar magnitude can be defined also for the case of finite, simply connected covering surfaces over the Riemann sphere. Here we assume that the covering surface has an analytic boundary.[6] Indeed, let F^* be a surface obtained from such a surface F by means of the stereographic mapping to the plane. If the boundary ∂F^* of F^* is described by a function

$$\varphi = \varphi(z), \quad |z| = 1, \tag{2.2.1}$$

then the magnitude $\nu(F,a)$ is defined similarly to $\nu(r,a)$ for $r = 1$.

Now we turn to statements of the results of Section 2.1.

Theorem 1. *Let $w(z) \in M_R$. Then for any point $a \in \overline{\mathbb{C}}$ and any r $(0 < r < R, w(z) \neq a, |z| = r)$*

$$\nu(r,a) + n(r,a) = A(r) + hL(r), \tag{2.2.2}$$

where $|h| < h(a) = \text{const} < \infty$.

---

[6]The analyticity of ∂F is not a very restrictive condition since a "less smooth" boundary can be approximated by the analytic ones. Hence, one can obtain some results for more general boundaries by a little deformation of F and ∂F, leading to arbitrarily small changes of magnitudes which we consider below.

Theorem 2. *Let F be a finite, simply connected surface having analytic boundary, lying over the Riemann sphere. Then for any $a \in \overline{\mathbb{C}}$, $a \notin \partial F$,*

$$\nu(F, a) + n(F, a) = A(F) + hL(F), \qquad (2.2.3)$$

where $|h| < h(a) = \text{const} < \infty$.

Using Ahlfors' Second Fundamental Theorem one can obtain its following analogs.

Theorem 3. *Let $w(z) \in M_R$. Then for any collection of pairwise different points $a_i \in \overline{\mathbb{C}}$, $i = 1, 2, \ldots, q$, and any $r \in (0, R)$*

$$\sum_{i=1}^{q} \nu(r, a_i) + \sum_{i=1}^{q} n_1(r, a_i) \leq 2A(r) + hL(r), \qquad (2.2.4)$$

where $|h| < h(a_1, a_2, \ldots, a_q) = \text{const} < \infty$.

The above theorem is a particular case of the following result which is a consequence of Theorem 2 and the second fundamental theorem of Ahlfors.

Theorem 4. *Let F be a finite, simply connected surface having analytic boundary, lying over the Riemann sphere. Then for any collection of pairwise different points $a_i \in \overline{\mathbb{C}}$, $i = 1, 2, \ldots, q$,*

$$\sum_{i=1}^{q} \nu(F, a_i) + \sum_{i=1}^{q} n_1(F, a_i) \leq 2A(F) + hL(F), \qquad (2.2.5)$$

where $|h| < h(a_1, a_2, \ldots, a_q) = \text{const} < \infty$.

Theorems 1–4 can be considered as some analogs of the First and Second Fundamental Theorems of Nevanlinna and Ahlfors. For the regularly exhaustible surface F (i.e. it is exhaustible by some surfaces $F_k \to F$ for which

$$\lim_{k \to \infty} \frac{L(F_k)}{A(F_k)} = 0, \Bigg)$$

and corresponding to this surface meromorphic function $w(z)$ we define by analogy the following deficiencies:

$$\overline{\delta}(a, F) = \lim_{k \to \infty} \inf \frac{\nu(F_k, a)}{A(F_k)} \qquad (2.2.6)$$

and

$$\overline{\delta}(a, w) = \lim_{r \to R} \inf \frac{\nu(r, a)}{A(r)}. \qquad (2.2.7)$$

Then the inequalities (2.2.4) and (2.2.5) imply the following analogs of deficiency relations:

$$\sum_{(i)} \overline{\delta}(a_i, F) \leq 2 \qquad (2.2.8)$$

and

$$\sum_{(i)} \overline{\delta}(a_i, w) \leq 2. \qquad (2.2.9)$$

Recall that the surfaces F_r is regularly exhaustible for any function given in $\mathbb{C}$ and also for any function given in $|z| < R < \infty$, such that $\lim_{r \to R} \sup(R - r)$ $A(r) = \infty$. Therefore for such functions (2.2.9) holds.

Theorems 1–4 represent definite statements on the structure of the boundary of F_r (or F) in the neighborhood of the point a for the wide class of regularly exhaustible surfaces and functions corresponding to them. Namely, if $w(z)$ takes the value a in $|z| \leq r$ (or on F) "rarely", then the closeness of some parts of the boundary of F_r (or F) to the point a is realized in a definite way: the boundary arcs are strongly coiled around the point a.

Note that in the definition of the magnitude $\nu(r, a)$, analogously to the Nevanlinna's approximation function $m(r, a)$, we did not use the closeness of the boundary arcs ∂F_r to a. However, this definition qualitatively contains such a conclusion: "a great number" of turns around a, together with "small" length of ∂F_r (because of the condition of regular exhaustibility) means that the arcs ∂F_r must be compressed around a.

Theorem 2, considering values $a \in D$ instead of domains D, fills the above mentioned gap (the absence of the First Fundamental Theorem) in Ahlfors' theory.

Theorems 2 and 4 contain some statements from differential geometry, see the author's paper [1].

2.2.3. The proof of Theorem 1 follows from two lemmas below.

Lemma 1. *Let* $w(z) \in M_R$. *Then for any* $a \in \overline{\mathbb{C}}$ *and any* $r \in (0, R)$

$$n(r, a) - A(r) = \frac{r}{2\pi} \int_{\Delta_1(r,a)} \frac{\partial}{\partial t} \ln |w(z) - a| d\varphi + hL(r), \quad a \neq \infty \ (2.2.10)$$

and

$$n(r,\infty) - A(r) = -\frac{r}{2\pi} \int_{\Delta_2(r,\infty)} \frac{\partial}{\partial t} \ln|w(z)| d\varphi + hL(r), \quad a = \infty, (2.2.11)$$

where $\Delta_1(r,a) = \{z : |z| = r, \ |w(z) - a| < 1\}$, $\Delta_2(r,\infty) = \{z : |z| = r, \ |w(z)| \geq 1\}$ *and* $|h| < h_1(a) = \text{const} < \infty$.

The initial point for proving (10) is the following identity (see Hayman [2])

$$n(r,a) - A(r) = \frac{r}{2\pi} \int_{|z|=r} \frac{\partial}{\partial r} \ln \frac{|w(z) - a|}{\sqrt{1 + |w(z)|^2}} d\varphi. \qquad (2.2.12)$$

For deriving this identity we need the following particular case of Green's formula, sometimes called the formula of Gauss. Let the function $u = u(x,y)$ be twice continuously differentiable by x and y in a simply connected domain G with piecewise smooth boundary ∂G. If n is the exterior normal to ∂G, Δ is the Laplace operator. ds is the length element of ∂G and $d\sigma$ is the area element, then

$$\int_{\partial G} \frac{\partial u}{\partial n} ds = \int \int_G \Delta u d\sigma.$$

We surround the poles b_ν. $|b_\nu| < r$, of the function $w(z)$ by no overlapping disks $D(b_\nu, \rho)$ of the radii ρ and draw cuts γ_ν from each disk to the circle $|z| = r$. Further, we set $u = \ln\sqrt{1 + |w(z)|^2}$ and as the domain $G = G_\rho$ we take what we get from $|z| < r$ after removing the disks $D(b_\nu, \rho)$ and the cuts γ_ν.

Observe that moving around our G_ρ in the positive direction we twice pass along the cuts γ_ν, having opposite normal. On the other hand, the exterior normal to the boundary of G_ρ is an interior one for $D(b_\nu, \rho)$ when moving around a disk $D(b_\nu, \rho)$. Therefore

$$\int_{\partial G_\rho} \frac{\partial}{\partial n} \ln\sqrt{1 + |w(z)|^2} ds$$

$$= \frac{1}{2\pi} \int_{|z|=r} \frac{\partial}{\partial r} \ln\sqrt{1 + |w(re^{i\varphi})|^2} r d\varphi$$

$$- \sum_{(\nu)} \frac{1}{2\pi} \int_{|z-b_\nu|=\rho} \frac{\partial}{\partial \rho} \ln\sqrt{1 + |w(b_\nu + \rho e^{i\vartheta})|^2} \rho d\vartheta$$

$$= \frac{1}{2\pi} \int \int_{G_\rho} \Delta \ln\sqrt{1 + |w(z)|^2} d\sigma. \qquad (2.2.13)$$

If k_ν is the multiplicity of the pole b_ν, then $w(z) = c_\nu(z - b_\nu)^{-k_\nu} + \cdots$ in a neighborhood of b_ν. Therefore letting $\rho \to 0$ on the circles $|z - b_\nu| = \rho$ we have

$$|w(b_\nu + \rho e^{i\vartheta})| = \frac{|c_\nu|}{\rho^{k_\nu}} + O(1)$$

and

$$|w(b_\nu + \rho e^{i\vartheta})|'_\rho = -\frac{|c_\nu| k_\nu}{\rho^{k_\nu+1}} + O(1),$$

and consequently

$$\frac{\partial}{\partial \rho} \ln \sqrt{1 + |w(b_\nu + \rho e^{i\vartheta})|^2}$$

$$= \frac{|w(b_\nu + \rho e^{i\vartheta})|'_\rho |w(b_\nu + \rho e^{i\vartheta})|}{1 + |w(b_\nu + \rho e^{i\vartheta})|^2} = -\frac{k_\nu}{\rho} + O(1).$$

Hence from (2.2.13) we get

$$\frac{1}{2\pi} \int_{|z|=r} \frac{\partial}{\partial r} \ln \sqrt{1 + |w(re^{i\varphi})|^2} r d\varphi + n(r, \infty)$$

$$= \lim_{\rho \to 0} \frac{1}{2\pi} \iint_{G_\rho} \Delta \ln \sqrt{1 + |w(z)|^2} d\sigma. \qquad (2.2.14)$$

Now we shall show that the last limit is equal to $A(r)$. Using the relation $|w|^2 = w\overline{w}$ we simplify the necessary calculations and get

$$\Delta \ln \sqrt{1 + |w(z)|^2} := \frac{1}{2} \left(\frac{(w\overline{w})'_x}{1 + |w|^2} \right)'_x + \frac{1}{2} \left(\frac{(w\overline{w})'_y}{1 + |w|^2} \right)'_y$$

$$= \frac{1}{2(1 + |w|^2)^2} \{I_1 + I_2\}. \qquad (2.2.15)$$

where

$$I_1 = (w'_x \overline{w} + w\overline{w}'_x)'_x(1 + |w|^2) - ((w\overline{w})'_x)^2,$$

$$I_2 = (w'_y \overline{w} + w\overline{w}'_y)'_y(1 + |w|^2) - ((w\overline{w})'_y)^2.$$

The expression in the figure brackets is equal to

$$(w''_{xx}\overline{w} + 2w'_x\overline{w}'_x + w\overline{w}''_{xx})(1 + |w|^2) - ((w\overline{w})'_x)^2$$

$$+ (w''_{yy}\overline{w} + 2w'_y\overline{w}'_y + w\overline{w}''_{yy})(1 + |w|^2) - ((w\overline{w})'_y)^2. \qquad (2.2.16)$$

On the other hand, by the Cauchy–Riemann conditions

$$w''_{xx} + w''_{yy} = 0, \quad \overline{w}''_{xx} + \overline{w}''_{yy} = 0, \quad w'_x \overline{w}'_x = w'_y \overline{w}'_y = |w'|^2,$$

$$((w\overline{w})'_x)^2 + ((w\overline{w})'_y)^2 = 4|w|^2|w'|^2.$$

Therefore the expression (2.2.16) is equal to $4|w'|^2$, and from (2.2.15) we get

$$\Delta \ln \sqrt{1 + |w(z)|^2} = \frac{2|w'(z)|^2}{(1 + |w(z)|^2)^2}.$$

Thus, we conclude that

$$\lim_{\rho \to 0} \frac{1}{2\pi} \iint_{G_\rho} \Delta \ln \sqrt{1 + |w(z)|^2} d\sigma$$

$$= \lim_{\rho \to 0} \frac{1}{\pi} \iint_{G_\rho} \frac{|w'(z)|^2}{(1 + |w(z)|^2)^2} d\sigma$$

$$:= \frac{1}{\pi} \iint_{D_r} \frac{|w'(z)|^2}{(1 + |w(z)|^2)^2} d\sigma := A(r). \qquad (2.2.17)$$

Now (2.2.13), (2.2.14) and (2.2.17) imply

$$n(r, \infty) - A(r) = \frac{r}{2\pi} \int_{|z|=r} \frac{\partial}{\partial r} \ln \frac{1}{\sqrt{1 + |w(re^{i\varphi})|^2}} d\varphi. \qquad (2.2.18)$$

On the other hand, by the argument principle

$$n(r, a) - n(r, \infty) = \frac{r}{2\pi} \int_{|z|=r} d\arg(w(z) - a) = \frac{r}{2\pi} \int_{|z|=r} \frac{\partial}{\partial r} \ln |w(z) - a| d\varphi.$$

Adding the last two equalities we come to (2.2.12).

For obtaining (2.2.10) now it remains to show that

$$\frac{r}{2\pi} \int_{|z|=r} \frac{\partial}{\partial r} \ln \frac{|w(z) - a|}{\sqrt{1 + |w(z)|^2}} d\varphi = \frac{r}{2\pi} \int_{\Delta_1(r,a)} \frac{\partial}{\partial r} \ln |w(z) - a| d\varphi + hL(r).$$

$$(2.2.19)$$

We have

$$\frac{r}{2\pi} \int_{|z|=r} \frac{\partial}{\partial r} \ln \frac{|w(z) - a|}{\sqrt{1 + |w(z)|^2}} d\varphi$$

$$= \frac{r}{2\pi} \int_{\Delta_1(r,a)} \frac{\partial}{\partial r} \ln |w(z) - a| d\varphi$$

$$- \frac{r}{2\pi} \int_{\Delta_1(r,a)} \frac{\partial}{\partial r} \ln \sqrt{1 + |w(z)|^2} d\varphi$$

$$+ \frac{r}{2\pi} \int_{\Delta_2(r,a)} \frac{\partial}{\partial r} \ln \frac{|w(z) - a|}{\sqrt{1 + |w(z)|^2}} d\varphi. \qquad (2.2.20)$$

Taking into account that $||w|_r'| \leq |w'|$ and that $|w(z)| < 1 + |a|$ for $z \in \Delta_1(r, a)$ and that due to the definition (see Subsection 2.1.9)

$$L(r) := \int_{|z|=r} \frac{|w'(z)|}{1 + |w(z)|^2} r d\varphi$$

we get

$$\left| \frac{r}{2\pi} \int_{\Delta_1(r,a)} \frac{\partial}{\partial r} \ln \sqrt{1 + |w(z)|^2} d\varphi \right| \leq \frac{r}{2\pi} \int_{\Delta_1(r,a)} \frac{||w(z)|_r'| |w(z)|}{1 + |w(z)|^2} d\varphi$$

$$\leq \frac{1 + |a|}{2\pi} \int_{\Delta_1(r,a)} \frac{|w'(z)|}{1 + |w(z)|^2} r d\varphi$$

$$\leq \frac{1 + |a|}{2\pi} \int_{|z|=r} \frac{|w'(z)|}{1 + |w(z)|^2} r d\varphi$$

$$= \frac{1 + |a|}{2\pi} L(r). \qquad (2.2.21)$$

Further,

$$\left| \left(\ln \frac{|w - a|}{\sqrt{1 + |w|^2}} \right)_r' \right| = \frac{1}{2} \left| \left(\ln \frac{|w - a|^2}{1 + |w|^2} \right)_r' \right|$$

$$= \frac{1}{2} \left| \frac{[(w - a)(\overline{w} - \overline{a})]_r'}{|w - a|^2} - \frac{(w\overline{w})_r'}{1 + |w|^2} \right|$$

$$= \frac{1}{2} \left| \frac{(w\overline{w})_r' - w_r' \overline{a} - \overline{w}_r' a}{|w - a|^2} - \frac{(w\overline{w})_r'}{1 + |w|^2} \right|$$

$$= \frac{1}{2} \left| \frac{(w\overline{w})_r'(1 + |w|^2 - |w - a|^2)}{|w - a|^2(1 + |w|^2)} - \frac{w_r' \overline{a} + \overline{w}_r' a}{|w - a|^2} \right|. \qquad (2.2.22)$$

As

$$||w|^2 - |w - a|^2| = |w\overline{w} - (w - a)(\overline{w} - \overline{a})|$$

$$= |\overline{w}a + w\overline{a} - |a|^2| \leq 2|w||a| + |a|^2,$$

$$|w_r'\bar{a} + \bar{w}_r'a| \le 2|a||w'|, \quad |(w\bar{w})_r'| \le 2|w||w'|,$$

from (2.2.22) we obtain

$$\left|\left(\ln\frac{|w-a|}{\sqrt{1+|w|^2}}\right)_r'\right| \le \frac{|w'||w|\left(1+2|w||a|+|a|^2\right)}{|w-a|^2\left(1+|w|^2\right)} + \frac{|w'||a|}{|w-a|^2}$$

$$= \frac{3|w|^2|a|+(1+|a|^2)|w|+|a|}{|w-a|^2}\frac{|w'|}{1+|w|^2}.$$

If $z \in \Delta_2(r,a)$, then the right-hand side expression of this relation is less than $h_2(a)|w'|(1+|w|^2)^{-1}$, where $h_2(a) = \text{const} < \infty$. Hence

$$\left|\frac{r}{2\pi}\int_{\Delta_2(r,a)}\frac{\partial}{\partial r}\ln\frac{|w(z)-a|}{\sqrt{1+|w(z)|^2}}d\varphi\right|$$

$$\le \frac{h_2(a)}{2\pi}\int_{\Delta_2(r,a)}\frac{|w'(z)|}{1+|w(z)|^2}rd\varphi \le \frac{h_2(a)}{2\pi}L(r). \qquad (2.2.23)$$

Now from (2.2.20), (2.2.21) and (2.2.23) we derive (2.2.19) with $|h| < h_1(a) := (2\pi)^{-1}(1+|a|+h_2(a))$. Substituting (2.2.19) into (2.2.12) we come to (2.2.10).

For the proof of (2.2.11) we represent (2.2.18) in the form

$$n(r,\infty) - A(r) = \frac{r}{2\pi}\int_{\Delta_2(r,0)}\frac{\partial}{\partial r}\ln\frac{1}{|w(z)|}d\varphi$$

$$+ \frac{r}{2\pi}\int_{\Delta_1(r,0)}\frac{\partial}{\partial r}\ln\frac{1}{\sqrt{1+|w(z)|^2}}d\varphi$$

$$+ \frac{r}{2\pi}\int_{\Delta_2(r,0)}\frac{\partial}{\partial r}\ln\frac{|w(z)|}{\sqrt{1+|w(z)|^2}}d\varphi$$

and apply to the last two integrals (2.2.21) and (2.2.23).

Lemma 2. *Under the conditions of Lemma 2 also the following relations are valid:*

$$\frac{r}{2\pi}\int_{\Delta_1(r,a)}\frac{\partial}{\partial r}\ln|w(z)-a|d\varphi = -\nu(r,a) + hL(r), \qquad (2.2.24)$$

$$\frac{r}{2\pi}\int_{\Delta_2(r,\infty)}\frac{\partial}{\partial r}\ln|w(z)|d\varphi = \nu(r,\infty) + hL(r), \qquad (2.2.25)$$

where $|h| < h_3(a) = \text{const} < \infty$.

Proof. We say that the positive movement along the arc γ_a is that corresponding to the positive turn around the part $w^{-1}(\gamma_a)$ of the disk $|z| = r$. By this we fixed the initial point and the endpoint of the arc γ_a. Denote the increase of the function f on a set T by $\Delta_T f$. Further, suppose that one can single out an arc γ_a^0 from a given arc γ_a, such that

(i) the initial points of γ_a and γ_a^0 coincide,
(ii) $\Delta_{w^{-1}(\gamma_a^0)} \arg(w(z) - a) = 2\pi \, [\nu_{\gamma_a}]'$.
Then due to the definition, we have

$$\Delta_{w^{-1}(\gamma \backslash \gamma_a^0)} \arg(w(z) - a) < 2\pi$$

if $\gamma_a \backslash \gamma_a^0 \neq \emptyset$. Let us single out all arcs of type γ_a^0 from all arcs of type γ_a. Then by (ii) we have

$$\left| \Delta_{w^{-1}(\tilde{\gamma}_a^0)} \arg(w(z) - a) \right| < 2\pi, \qquad (2.2.26)$$

for the remaining parts of γ_a, which are the arcs $\tilde{\gamma}_a$ forming the set $\bigcup \gamma_a \backslash \bigcup \gamma_a^0$. Further, we single out from the arcs $\tilde{\gamma}_a^0$ those γ_a' the lengths of which are $l_a' \geq 1/2$, and those γ_a'', the lengths of which are $l_a'' < 1/2$. As this is a complete partition,

$$
\begin{aligned}
\frac{r}{2\pi} \int_{\Delta_1(r,a)} \frac{\partial}{\partial r} \ln|w(z) - a| d\varphi &= \frac{1}{2\pi} \int_{\Delta_1(r,a)} d \, \arg(w(z) - a) \\
&= \sum_{(\gamma_a^0)} \frac{1}{2\pi} \int_{w^{-1}(\gamma_a^0)} d \, \arg(w(z) - a) \\
&+ \sum_{(\gamma_a')} \frac{1}{2\pi} \int_{w^{-1}(\gamma_a')} d \, \arg(w(z) - a) \\
&+ \sum_{(\gamma_a'')} \frac{1}{2\pi} \int_{w^{-1}(\gamma_a'')} d \, \arg(w(z) - a).
\end{aligned}
$$

$$(2.2.27)$$

It follows from the definitions of $\nu(r,a)$ and γ_a^0 that

$$\sum_{(\gamma_a^0)} \frac{1}{2\pi} \int_{w^{-1}(\gamma_a^0)} d \, \arg(w(z) - a) = \sum_{(\gamma_a^0)} [\nu_{(\gamma_a^0)}] = -\nu(r,a). \qquad (2.2.28)$$

Any arc γ'_a also is an arc of type $\tilde{\gamma}_a$ for which (2.2.26) is true; therefore,

$$\left| \sum_{(\gamma'_a)} \frac{1}{2\pi} \int_{w^{-1}(\gamma'_a)} d\, \arg(w(z) - a) \right| \leq \sum_{(\gamma'_a)} 1 \leq 2 \sum_{(\gamma'_a)} l'_a. \qquad (2.2.29)$$

Let

$$\left| \int_{w^{-1}(\gamma''_a)} d\, \arg(w(z) - a) \right| = 2\pi\alpha$$

for a given arc γ''_a. Because $l''_a < 1/2$, at least one of the endpoints of γ''_a is on the circle $|w - a| = 1$. Hence γ''_a is completely contained in the domain $1/2 < |w - a| < 1$ and is seen from the point a in an angle which is not less than $2\pi\alpha$. Therefore the length of l''_a is not less than the length of the arc from $|w - a| = 1/2$ lying in this angle, i.e. $l''_a > \pi\alpha$. Consequently

$$\left| \frac{1}{2\pi} \int_{w^{-1}(\gamma''_a)} d\, \arg(w(z) - a) \right| = \alpha < \frac{l''_a}{\pi}$$

and

$$\left| \sum_{(\gamma''_a)} \frac{1}{2\pi} \int_{w^{-1}(\gamma''_a)} d\, \arg(w(z) - a) \right| \leq \frac{1}{\pi} \sum_{(\gamma''_a)} l''_a. \qquad (2.2.30)$$

Substituting relations (2.2.28)–(2.2.30) into (2.2.27) we obtain

$$\frac{r}{2\pi} \int_{\Delta_1(r,a)} \frac{\partial}{\partial r} \ln|w(z) - a| d\varphi = -\nu(r, a) + hl, \qquad (2.2.31)$$

where l is the sum of lengths of all arcs γ'_a, γ''_a and $h < h_4(a) = \text{const} < \infty$. Evidently, the lengths of these arcs are comparable with the lengths of their spheric images as all these arcs are in the unit circle centered at the point a. Consequently, $l < h_5(a)L(r)$ for a constant $h_5(a) < \infty$, and taking into account (2.2.31) we come to (2.2.24).

For proving the relation (2.2.25) we write (2.2.24) for the function $W = 1/w$ and for $a = 0$. Then

$$\frac{r}{2\pi} \int_{\Delta_1(r,0,W)} \frac{\partial}{\partial r} \ln|W(z)| d\varphi = -\nu(r, 0, W) + hL(r, W). \qquad (2.2.32)$$

Clearly the left-hand side integral is equal to

$$-\frac{r}{2\pi} \int_{\Delta_2(r,0,w)} \frac{\partial}{\partial r} \ln |w(z)| d\varphi.$$

At the same time, $\nu(r, 0, W) = \nu(r, \infty, w)$ and $L(r, W) = L(r, w)$. Hence the relation (2.2.32) coincides with (2.2.25), and Lemma 2 is completely proved.

It is evident that Theorem 1 is a consequence of Lemmas 1 and 2.

Theorem 2 easily follows from Theorem 1. Indeed, by the main theorem of the theory of Riemann surfaces (see Stoilov [2]) there exists a function meromorphic in the unit disk $|z| < 1$ mapping this disk to the surface F. As the boundary of F is an analytic curve, $w(z)$ has an analytic continuation out of the unit disk (see, for instance, Courant [1]). Therefore

$$\nu(1, a, w) + n(1, a, w) = A(1, w) + hL(1, w).$$

Hence (2.2.3) follows by the obvious relations

$$n(1, a, w) = n(F, a), \quad \nu(1, a, w) = \nu(F, a),$$

$$A(1, w) = A(F), \quad L(1, w) = L(F).$$

2.2.4. Remark 1. From Theorem 1 for the cases $a \neq \infty$ and $a = \infty$ the following interpretation of the Argument Principle holds:

$$\frac{1}{2\pi} \int_{|z|=r} d \, \arg(w(z) - a) = n(r, a) - n(r, \infty) = \nu(r, \infty) - \nu(r, a) + hL(r).$$

On the other hand, this new form of the Argument Principle is actually an analog of the First Fundamental Theorem since the last equality can be written in the form

$$n(r, a) + \nu(r, a) = n(r, \infty) + \nu(r, \infty) + hL(r).$$

Remark 2. The following analog of Cartan's identity is a consequence of Lemma 1:

$$A(r) = \frac{1}{2\pi} \int_0^{2\pi} n(r, e^{i\vartheta}) d\vartheta + hL(r). \qquad (2.2.33)$$

Indeed, differentiating Cartan's identity (see Section 2.1.3) with respect to $\ln t$ we obtain the equality

$$\frac{r}{2\pi} \int_{\Delta_2(r,0)} \frac{\partial}{\partial r} \ln |w(z)| d\varphi + n(r,\infty) = \frac{1}{2\pi} \int_0^{2\pi} n(r, e^{i\vartheta}) d\vartheta.$$

Applying the inequality (2.2.11) of Lemma 1 to the left-hand side of this equality we find (2.2.33).

2.3. On some Additions to Ahlfors' Theory of Covering Surfaces

Ahlfors' Theory of Covering Surfaces offers another metric topological approach to the main conclusions of Nevanlinna's Value Distribution Theory and generalizes the last one if $w(z)$ is a meromorphic function in the complex plane. This generalization is achieved due to the circumstance that the numbers of islands (the main concept in Ahlfors' Theory) can be considered as a generalization of the numbers of a-points of the functions $w(z)$. However, until now the pre-images of the islands (which actually themselves are analogs of the a-points) were not studied. In this section, we give an addition to Ahlfors' Theory by the geometric study of pre-images of the islands and by the study of derivatives on the set of a-points of the function $w(z)$. Some connections between Ahlfors' Theory and Γ-lines' Theory are established as well.

2.3.1. Suppose that $w(z)$ is a meromorphic function in $\mathbb{C}$; $a_1, a_2, \ldots, a_q$ be pairwise different complex values; Ω_a be a subset of the set $\{z_i(a_\nu)\}$, $i = 1, 2, \ldots$, $\nu = 1, 2, \ldots, q$, where $z_i(a_\nu)$ are a_ν-points in $\{z : |z| < r\}$; $n(\Omega_a, a_\nu)$ be the number of a_ν-points (without counting multiplicities) belonging to Ω_a.

Definition 1. *We shall call the sets* Ω_a *Ahlfors' sets of simple a-points if*

$$\sum_{\nu=1}^{q} n(\Omega_a, a_\nu) \geq (q-4)A(r) - o(A(r)). \quad r \to \infty, \quad r \notin E, \qquad (2.3.1)$$

where E is a set of finite logarithmic measure.

Then due to the Second Fundamental Theorem of Ahlfors we have:

Theorem A. *Given a meromorphic function in $\mathbb{C}$ and pairwise different fixed values $a_1, a_2, \ldots, a_q$, there exists Ahlfors' sets Ω_a of simple a-points in the disks $\{z : |z| < r\}$, $r \notin E$.*

Similarly, suppose that $D_1, D_2, \ldots, D_q$ are simple connected domains with non overlapping closures in the complex plane such that the distances between these domains are positive; Ω_D be a subset of the set of the simple islands $\{(D_\nu^i)\}$, $i = 1, 2, \ldots$, $\nu = 1, 2, \ldots, q$ of the surface $F_r := \{w(z) : |z| < r\}$ over these domains D_ν; $n(\Omega_D, D_\nu)$ be the number of simple islands of the surface F_r over the domain D_ν belonging to Ω_D.

Similarly, suppose that $\Gamma_1, \Gamma_2, \ldots, \Gamma_q$ are smooth Jordan non-intersecting curves in the complex plane such that the distances between these curves are positive; Ω_Γ be a subset of the set of the simple islands $\{(\Gamma_\nu^i)\}$, $i = 1, 2, \ldots$, $\nu = 1, 2, \ldots, q$ of the surface F_r over these curves Γ_ν; $n(\Omega_\Gamma, \Gamma_\nu)$ be the number of simple islands of the surface F_r over the curve Γ_ν belonging to Ω_Γ.

Definition 1′. *We shall call the sets Ω_D Ahlfors' sets of simple islands of F_r over $D_1, D_2, \ldots, D_q$ if*

$$\sum_{\nu=1}^{q} n(\Omega_D, D_\nu) \geq (q - 4)A(r) - o(A(r)), \quad r \to \infty, \quad r \notin E, \qquad (2.3.2)$$

where E is a set of finite logarithmic measure.

Definition 1″. *We shall call the sets Ω_Γ Ahlfors' sets of simple islands of F_r over $\Gamma_1, \Gamma_2, \ldots, \Gamma_q$ if*

$$\sum_{\nu=1}^{q} n(\Omega_\Gamma, \Gamma_\nu) \geq (q - 4)A(r) - o(A(r)), \quad r \to \infty, \quad r \notin E, \qquad (2.3.3)$$

where E is a set of finite logarithmic measure.

Then due to the Second Fundamental Theorem of Ahlfors, the following results are valid.

Theorem A′. *Given a meromorphic function in $\mathbb{C}$ and the above domains $D_1, D_2, \ldots, D_q$ then there exists Ahlfors' sets Ω_D of simple islands of F_r over $D_1, D_2, \ldots, D_q$.*

Theorem A″. *Given a meromorphic function in $\mathbb{C}$ and the above curves $\Gamma_1, \Gamma_2, \ldots, \Gamma_q$ then there exists Ahlfors' sets Ω_G of simple islands of F_r over $\Gamma_1, \Gamma_2, \ldots, \Gamma_q$.*

Theorems A, A′ and A″ "almost" coincide with the main conclusions of Ahlfors' theory concerning the description of simple a-points, simple islands over domains and curves.

Below we give some additions to these results. In the case of Theorem A we give bounds for $|w'(z)|$ where z coincide with simple a-points appearing in the theorem (see the author's article [14]). In the case of Theorem A' we give bounds for diameters of pre-images of simple islands over domains appearing in the theorem. In the case of Theorem A'' we give bounds for the lengths of pre-images of simple islands over curves appearing in the theorem.

Theorem 1. *Let $w(z)$ be a meromorphic function in $\mathbb{C}$; $\varphi(r)$ be a monotone function increasing on $[0, \infty)$ and $\varphi(r) \to \infty$ as $r \to \infty$. Then for any pairwise different fixed values $a_1, a_2, \ldots, a_q$ there exists Ahlfors' sets Ω_a in the disks $\{z : |z| < r\}$, $r \notin E$, and moreover for any a_ν-point $z_k(a_\nu)$ belonging to Ω_a*

$$|w'(z_k(a_\nu))| \geq \frac{1}{\varphi(r)} \frac{A^{1/2}(r)}{r} \qquad (2.3.4)$$

holds.

Theorem 2. *Let $w(z)$ be a meromorphic function in $\mathbb{C}$; $\varphi(r)$ be a monotone function increasing on $[0, \infty)$ and $\varphi(r) \to \infty$ as $r \to \infty$. Then for the above defined domains $D_1, D_2, \ldots, D_q$ there exists Ahlfors' sets Ω_D of simple islands D_ν^k of the surface F_r over these domains D_ν such that for any ν and k*

$$d(w^{-1}(D_\nu^k)) \leq \frac{\varphi(r)r}{A^{1/2}(r)}, \qquad (2.3.5)$$

holds, where $d(X)$ is the diameter of X.

Theorem 3. *Let $w(z)$ be a meromorphic function in $\mathbb{C}$; $\varphi(r)$ be a monotone function increasing on $[0, \infty)$ and $\varphi(r) \to \infty$ as $r \to \infty$. Then for the above defined curves $\Gamma_1, \Gamma_2, \ldots, \Gamma_q$ there exists Ahlfors' sets Ω_Γ of simple islands Γ_ν^k on the surface F_r over these curves Γ_ν such that for any ν and k*

$$l(w^{-1}(\Gamma_\nu^k)) \leq \frac{\varphi(r)r}{A^{1/2}(r)} \qquad (2.3.6)$$

holds, where $l(X)$ is the length of X.

Remark 1. For arbitrary Ahlfors' sets Ω_Γ of simple islands Γ_ν^k of the surface F_r over the curves Γ_ν exist no more than

$$qA(r) + o[A(r)], \qquad r \to \infty, \quad r \notin E$$

curves (according to the Second Fundamental Theorem of Ahlfors). Therefore from (2.3.6) we obtain

$$\sum_{\nu}\sum_{k} l(w^{-1}(\Gamma_\nu^k)) \leq (q+1)\varphi(r)rA^{1/2}(r), r \notin E. \tag{2.3.7}$$

We shall show in Section 4.1 that with an absolute constant K the inequality

$$\sum_{\nu} L(r,\Gamma_\nu) \leq KrA(r)$$

holds for a sequence $r_n \to \infty$ and that this is an exact estimate in general. Hence, (2.3.7) shows that the part of Γ_ν-lines involved in w^{-1}-images of Ahlfors' sets of simple islands has much less length than the length of all Γ_ν-lines in the general case.

Remark 2. In the case of meromorphic functions in $\{z:|z| < R < \infty\}$ satisfying (2.1.30) in Section 2.1 one can establish some results similar to Theorems 1, 2 and 3 with the only difference that instead of $r \to \infty$, $r \notin E$ appears in the sequence $r_n \to R$.

2.3.2. Proofs. Let Γ be a bounded curve. Suppose Γ belongs to a simply connected domain D which, in turn, belongs to a simply connected domain D', and assume that the spheric distance l_0 from ∂D to $\partial D'$ and the ordinary distance $\rho(\Gamma, D)$ from Γ to ∂D are positive. Suppose that $E_k(r)$, $k = 1, 2, \ldots, \Phi(r)$, are some domains in the z plane, such that

(i) the set $w(E_k(r))$ coincides with D,
(ii) $E_k(r) \bigcap D(r) \neq \emptyset$,
(iii) any $E_k(r)$ is contained in a domain $E'_k(r)$ where the function $w(z)$ is univalent, and $w(E'_k(r))$ coincides with D'.

In other words, if F is the whole surface generated by $w(z)$ or, which is the same, the Riemann surface of the function w^{-1}, then $w(E'_k(r))$ are the simple islands of F lying over D', $E'_k(r)$ are the pre-images of these simple islands, $E_k(r)$ are the parts of $E'_k(r)$ such that $E_k(r)$ have common points with $D(r)$.

Further, we denote the sets $w^{-1}(\Gamma) \bigcap E_k(r)$ by l_k^Γ, $k = 1, 2, \ldots, \Phi(r)$ (obviously it is only one curve), and the length of a curve X by $|X|$. Evidently, if the Riemann surface F_r is not ramified over the domain D', then the totality of Γ-lines consists of only sets l_k^Γ so that

$$L(r.\Gamma) = \sum_{k=1}^{\Phi(r)} \left| l_k^\Gamma \right|. \tag{2.3.8}$$

To prove the above theorems we first establish some preliminary assertions. Some of them are of independent interest.

Lemma 1. *There exist some constants K_1 and K_2 depending on D, D' and Γ, such that*

$$K_1 d(E_k(r)) \leq \left| l_k^\Gamma \right| \leq K_2 d(E_k(r)) \tag{2.3.9}$$

and there exist some constants K_3 and K_4 depending only on D, D' such that for any point $z \in E_k(r)$

$$K_3 d(E_k(r)) \leq \frac{1}{|w'(z)|} \leq K_4 d(E_k(r)). \tag{2.3.9'}$$

Proof. First we shall show that there exist some constants $K_1(D, D')$ and $K_2(D, D')$ depending only on the distance $\rho(\partial D, \partial D')$ and the geometry of the domains D and D'. such that for any two points w_1 and w_2 from the closure of $w(E_k(r))$[7]

$$K_1(D, D')|z_k'(w_2)| \leq |z_k'(w_1)| \leq K_2(D, D')|z_k'(w_2)|, \tag{2.3.10}$$

where $z_k(w)$ is the branch of the function w^{-1} defined in the domain $w(E_k(r))$.

We prove this by standard methods of the theory of univalent functions. Namely, in $w(E_k(r))$ we choose a path l connecting the points w_1 and w_2 such that l can be covered by some disks $s_i. i = 1, 2, \ldots. n_l$, of the radii $\rho(\partial D, \partial D')$, with the distance $\leq \rho(\partial D, \partial D')/2$ between the centers of neighboring disks. Further, we denote the centers of these disks by $w_i^*. i = 1, 2, \ldots, n_l$, where $w_1^* = w_1$ and $w_{n_l}^* = w_2$. Evidently. the function $z_k(w)$ gives a one-to-one mapping of the domain $w(E_k'(r))$ and hence of each disk s_i.

Now we make use of the following known Koebe distortion theorem: for a holomorphic one-to-one mapping $f(z) = f(z_0) + f'(z_0)(z - z_0) + \cdots$ in the disk $|z - z_0| < \rho$

$$\frac{(\rho)^2|z - z_0||f'(z_0)|}{(\rho + |z - z_0|)^2} \leq |f(z) - f(z_0)| \leq \frac{(\rho)^2|z - z_0||f'(z_0)|}{(\rho - |z - z_0|)^2} \tag{2.3.11}$$

---

[7]By definition, a domain $w(E_k(r))$ is a univalent domain lying over $D \subset D'$.

holds and

$$\frac{(\rho)^2(\rho - |z - z_0|)|f'(z_0)|}{(\rho + |z - z_0|)^3} \leq |f'(z)| \leq \frac{(\rho)^2(\rho + |z - z_0|)|f'(z_0)|}{(\rho - |z - z_0|)^3}. \quad (2.3.12)$$

Therefore, applying the inequality (2.3.12) to $z_k(w)$ we obtain

$$\frac{4}{27}\left|z'_k(w^*_{i+1})\right| \leq \left|z'_k(w^*_i)\right| \leq 12\left|z'_k(w^*_{i+1})\right|.$$

Repeating this procedure for all the disks we get

$$\left(\frac{4}{27}\right)^{n_l-1}\left|z'_k(w^*_{i+1})\right| \leq \left|z'_k(w^*_i)\right| \leq 12^{n_l-1}\left|z'_k(w^*_{i+1})\right|.$$

Obviously, there exists a number n^* depending on D and D', such that any two points can be connected, as above, by a curve covered by not more than n^* disks. Hence the last inequality implies (2.3.10), where

$$K_1(D, D') = \left(\frac{4}{27}\right)^{n^*-1} \quad \text{and} \quad K_2(D, D') = 12^{n^*-1}.$$

Now let $\tilde{w}_k$ be a point in the domain $w(E_k(r))$, which is projected to $\tilde{w}$. Then by (2.3.10)

$$\begin{aligned} K_1(D, D')|\Gamma||z'_k(w_k(0))| &\leq |l^G_k| := \int_G |z'_k(w)|ds \\ &\leq K_2(D, D')|\Gamma||z'_k(w_k(0))|. \quad (2.3.13) \end{aligned}$$

On the other hand, assuming that $\rho(\tilde{w}, \partial D)$ is the distance from $\tilde{w}$ to the boundary of D, and that w is a point on the circle $|w - \tilde{w}_k| = \rho(\tilde{w}, \partial D)/2$, by (2.3.11) and (2.3.12) we derive

$$\begin{aligned} \frac{2}{9}\rho(\tilde{w}, \partial D)|z'_k(\tilde{w}_k)| &\leq |z_k(w) - z_k(\tilde{w})| \\ &\leq d(E_k(r)) \leq \int_{\partial D} |z'_k(w)|ds \\ &\leq K_2(D, D')|\partial D||z'_k(\tilde{w}_k)|. \quad (2.3.14) \end{aligned}$$

From (2.3.13) and (2.3.14) it follows that

$$\frac{K_1(D, D')|\Gamma|}{K_2(D, D')|\partial D|}d(E_k(r)) \leq |l^\Gamma_k| \leq \frac{9}{2\rho(\tilde{w}, \partial D)}K_2(D, D')|\Gamma|d(E_k(r)),$$

whence the inequality (2.3.9) of Lemma 1 follows. Inequality (2.3.9′) of Lemma 1 follows from (2.3.14) taking into account that $|w'(z)| = 1/|z'(w)|$.

Lemma 2. *If $w(z) \in M_R$, then there exists a constant K depending on D and D', such that for $r < R$*

$$\sum_{k=1}^{\Phi(r)} d(E_k(r)) \leq \frac{1}{l_0} \int_0^r L(t)dt + Kr \leq \frac{\pi}{l_0} r A^{1/2}(r) + Kr, \qquad (2.3.15)$$

where $d(E_k(r))$ is the diameter of $E_k(r)$.

For proving Lemma 2 we divide the domains $E_k(r)$, $k = 1, 2, \ldots, \Phi(r)$ to three types:

(a) the domains $E_k(1, r)$, $k = 1, 2, \ldots, n_0(r, D')$, for which the extended domains $E_k'(r)$ are contained in the open disk $D(r)$;

(b) the domains $E_k(2, r)$, $k = 1, 2, \ldots, n_0'(r, D)$, contained in $D(r)$ but such that $E_k'(r)$ has common points with the circle $|z| = r$;

(c) the domains $E_k(3, r)$, $k = 1, 2, \ldots, n_0''(r, D)$, having common points with $|z| = r$.

Obviously

$$\sum_{k=1}^{\Phi(r)} d(E_k(r)) = \sum_{k=1}^{n_0(r,D')} d(E_k(1, r))$$

$$+ \sum_{k=1}^{n_0'(r,D)} d(E_k(2, r)) + \sum_{k=1}^{n_0''(r,D)} d(E_k(3, r)). \quad (2.3.16)$$

First we show that

$$\sum_{k=1}^{n_0(r,D')} d(E_k(1, r)) \leq \frac{1}{l_0} \int_0^r L(t)dt + 4r. \qquad (2.3.17)$$

Let $\Phi_1(x, r)$ and $\Phi_1(y, r)$ be the numbers of domains $E_k(1, r)$ having common points with the intersections $J_x = \{z : \operatorname{Re} z = x\} \cap D(r)\}$ and $J_y = \{z : \operatorname{Im} z = y\} \cap D(r)\}$, respectively. Each of $E_k(1, r)$ is contained in a domain $E_k'(1, r)$ defined by the condition that $E_k(1, r) \subset E_k'(1, r)$ and $w(E_k'(1, r))$ univalently cover D'. Suppose that for a given x_0 holds $\Phi_1(x_0, r) \geq 2$. If $E_k(1, r)$ has common pointswith J_{x_0} then we can find two intervals

$\delta^*(1,k)$ and $\delta^{**}(1,k)$ on J_{x_0} such that each of them connects a point of the boundary of $\partial E_k(1,r)$ with a point of the boundary of $\partial E_k'(1,r)$. Consequently, w-images of these intervals have the spheric lengths $\rho(\delta^*(1,k))$ and $\rho(\delta^{**}(1,k))$ which are not less than l_0. Therefore, denoting the spheric lengths of the curves $w(J_x)$ and $w(J_y)$ by L_x and L_y respectively, we have

$$\Phi_1(x_0,r) \leq \frac{1}{2} \sum_{\lambda=1}^{\Phi_1(x_0,r)} \frac{\rho(\delta^*(1,k)) + \rho(\delta^{**}(1,k))}{l_0} \leq \frac{L_{x_0}}{2l_0}.$$

Taking into account also the case when $\Phi_1(x_0,r) \leq 1$ we obtain for all x

$$\Phi_1(x_0,r) \leq \frac{L_x}{2l_0} + 1, \qquad (2.3.18)$$

and similarly

$$\Phi_1(y_0,r) \leq \frac{L_y}{2l_0} + 1. \qquad (2.3.18')$$

Evidently the following integrals

$$\int_{-r}^{r} \Phi_1(x,r)dx, \quad \int_{-r}^{r} \Phi_1(y,r)dy$$

are equal respectively to

$$\sum_{k=1}^{n_0(r,D')} \left\{ \sup_{z \in E_1(x,r)} \operatorname{Re} z - \inf_{z \in E_1(x,r)} \operatorname{Re} z \right\}$$

and

$$\sum_{k=1}^{n_0(r,D')} \left\{ \sup_{z \in E_1(x,r)} \operatorname{Im} z - \inf_{z \in E_1(x,r)} \operatorname{Im} z \right\},$$

Since the diameters $dE_k(1,r)$ are less than or equal to the sum of the previous two magnitudes we obtain

$$\sum_{k=1}^{n_0(r,D')} d(E_k(1,r)) \leq \int_{-r}^{r} \Phi_1(x,r)dx + \int_{-r}^{r} \Phi_1(y,r)dy,$$

so that applying (2.3.18) and (2.3.18′) we get

$$
\sum_{k=1}^{n_0(r,D')} d(E_k(1,r)) \leq \frac{1}{2l_0}\left\{\int_{-r}^{r} L_x dx + \int_{-r}^{r} L_y dy\right\} + 4r
$$

$$
= \frac{1}{2l_0}\left\{\int_{-r}^{r}\left\{\int_{J_x}\frac{|w'|}{1+|w|^2 dy}\right\}dx\right.
$$

$$
\left. + \int_{-r}^{r}\left\{\int_{J_y}\frac{|w'|}{1+|w|^2 dx}\right\}dy\right\} + 4r
$$

$$
= \frac{1}{l_0}\int_{-r}^{r}\int_{0}^{2\pi}\frac{|w'|}{1+|w|^2}rdrd\varphi + 4r,
$$

so that applying the Cauchy–Buniacovski inequality we obtain the part of (2.3.15) corresponding to the domains $E_k(1,r)$.

Let now $\Phi_2(x,r)$ and $\Phi_2(y,r)$ be the numbers of domains $E_k(2,r)$ having common points with the intersections $J_x = \{z : \operatorname{Re} z = x\}\cap D(r)$ and $J_y = \{z : \operatorname{Im} z = y\}\cap D(r)$ respectively. Each of $E_k(2,r)$ is contained in a domain $E_k'(2,r)$ defined by the conditions that $E_k(2,r) \subset E_k'(2,r)$ and $w(E_k'(2,r))$ univalently cover D'. It can happen that for a given $E_k'(2,r)$ and a fixed x_0 at least one of the points of the set $\partial E_k'(2,r)\cap\{z : \operatorname{Re} z = x\}$ is out of $D(r)$. If $\Phi_2'(x_0,r)$ is the number of such domains. then obviously

$$
\Phi_2'(x_0,r) \leq 2.
$$

Consider now the remaining domains $E_\lambda(2,r)$. $\lambda = 1, 2, \ldots$, having common points with J_{x_0}. Obviously for any such λ there exist two intervals $\delta^*(2,\lambda)$ and $\delta^{**}(2,\lambda)$ each connecting a point of the boundary $\partial E_k(2,r)$ with a point of the boundary $\partial E_k'(2,r)$. Hence the w-images of these intervals have the spheric lengths $\rho(\delta^*(2,\lambda))$ and $\delta^{**}(2,\lambda)$ which are not less than l_0. Therefore, denoting the spheric lengths of the curves $w(J_x)$ and $w(J_y)$ by L_x and L_y respectively. we have

$$
\Phi_2(x_0,r) - \Phi_2'(x_0,r) \leq \frac{1}{2}\sum_{\lambda=1}^{\Phi_2(x_0,r)-\Phi_2'(x_0,r)}\frac{\rho(\delta^*(2,\lambda)) + \rho(\delta^{**}(2,\lambda))}{l_0} \leq \frac{L_x}{2l_0}.
$$

The last two inequalities imply

$$
\Phi_2(x,r) \leq \frac{L_x}{2l_0} + 2.
$$

In the same way we find

$$\Phi_2(y, r) \leq \frac{L_y}{2l_0} + 2.$$

Now similar to the proof of (2.3.15) we obtain

$$\sum_{k=1}^{n_0'(r,D)} d(E_k(2,r)) \leq \frac{1}{l_0} \int_0^r \int_0^{2\pi} \frac{|w'(z)|}{1 + |w'(z)|^2} r \, dr \, d\varphi + 8r. \qquad (2.3.19)$$

For evaluating the diameters of the domains $E_k(3,r)$ we construct a domain D'' containing D and contained in D' and suppose that the distances l_0'' from ∂D to $\partial D''$ and l_0' from $\partial D''$ to $\partial D'$ both are positive. By the definition, any domain $E_k(3,r)$ is intersected with the circle $|z| = r$. Hence for any $E_k(3,r)$, there is an arc $\delta_k \subset \{|z| = r\}$ connecting two points belonging to $\partial(E_k(3,r)) \cap \{|z| = r\}$ and $\partial(E_k''(3,r)) \cap \{|z| = r\}$ respectively, where $E_k'(3,r)$ is defined similarly as $E_k''(1,r)$. Consequently, the curve $w(\delta_k)$ connects ∂D with $\partial D''$, and therefore, $|w(\delta_k)| \geq l_0''$ so that denoting by z^* a point in δ_k, where $|w'(z)|$ takes its maximal value at δ_k, we obtain

$$|w'(z^*)||\delta_k| \geq |w(\delta_k)| := \int_{\delta_k} |w'(z)| r \, d\varphi \geq l_0''. \qquad (2.3.20)$$

As $l_0' > 0$, we can use the inequality (2.3.10) for the domains D'' and D'. Applying (2.3.10) for the points $w^* = w(z^*)$ and $\tilde{w}_k$ we get

$$K_1(D'', D')|z_k'(\tilde{w}_k)| \leq |z_k'(w^*)|,$$

whence by (2.3.20)

$$|z_k'(\tilde{w}_k)| \leq \frac{|z_k'(w^*)|}{K_1(D'', D')} = \frac{1}{K_1(D'', D')|w'(z^*)|} \leq \frac{|\delta_k|}{l_0'' K_1(D'', D')},$$

where $|\delta_k|$ is the length of δ_k. Taking into account also the inequality (2.3.14) and that $\sum \delta_k \leq 2\pi r$ we come to

$$\sum_{k=1}^{n_0''(r,D)} d(E_k(3,r)) \leq \frac{K_2(D, D')|\partial D|}{l_0'' K_1(D'', D')} \sum_{k=1}^{n_0''(r,D)} |\delta_k|$$

$$= K_3 \sum_{k=1}^{n_0''(r,D)} |\delta_k| \leq 2\pi K_3 r. \qquad (2.3.21)$$

Now the assertion of Lemma 2 follows with $K = 2\pi K_3 + 12$ from (2.3.16), (2.3.17), (2.3.19) and (2.3.21).

Now we prove Theorem 2. We make use of Lemmas 1 and 2 to our domains D_ν. We define some domains D'_ν similarly as in Lemma 2 so that the minimal distance between different D'_ν are positive. Denote by $E_k(r, D_\nu)$ the domains $E_k(r)$ in the proof of Lemma 2 applied to D_ν and by $E'_k(r, D_\nu)$ the domains $E'_k(r)$ in the proof of Lemma 2 applied to D_ν. Let $\overline{\Omega}_D$ be the totality of those domains $\overline{E}_k(r, D_\nu)$ of the type $E_k(r, D_\nu)$, taken for all k and ν for which every $\overline{E}_k(r, D_\nu)$ lies in $E'_k(r, D_\nu)$ which in turn lies completely in $\{z : |z| < r\}$. Obviously $n(\Omega_D, D_\nu) = n_0(r, D'_\nu)$ so that, clearly, $\overline{\Omega}_D$ is an Ahlfors' set of simple islands over $D_1, D_2, \ldots, D_q$. The w^{-1}-images of simple islands $\overline{D}^k_\nu \in \overline{\Omega}_D$ are our domains $\overline{E}_k(r, D_\nu)$. The numbers of these domains (which is equal to $n(r, D'_\nu)$) are less than $\Phi(r) := \Phi(r, D_\nu)$, so that from Lemma 2 it follows that

$$\sum_{k=1}^{n_0(r, D'_\nu)} d(\overline{E}_k(r, D_\nu)) \leq \frac{\pi}{l_0} r A^{1/2}(r) + Kr. \qquad (2.3.22)$$

Therefore, we obtain that the number of all islands $\overline{D}^k_\nu$ in $\overline{\Omega}_D$ for which

$$d(w^{-1}(\overline{D}^k_\nu)) \geq \frac{\varphi(r)r}{A^{1/2}(r)} \qquad (2.3.23)$$

is a magnitude $o[A(r)]$. Consequently by removing from $\overline{\Omega}_D$ all the simple islands $\overline{D}^k_\nu$ satisfying (2.3.23) we get a totality Ω_D of simple islands D^k_ν which is also an Ahlfors' set of simple islands over $D_1, D_2, \ldots, D_q$. But for these islands (2.3.5) already holds. Theorem 2 is proved.

To prove Theorem 1 we construct domains D_ν involving a_ν and D'_ν similarly as in Theorem 2. Let $z_k(a_\nu)$ be the a_ν-point belonging to $w^{-1}(D^k_\nu)$, $D^k_\nu \subset \Omega_D$. Since the number of such points $z_k(a_\nu)$ is not less than $n_0(r, D'_\nu) + o[A(r)]$ we have that the totality of these points $z_k(a_\nu)$ taken for all ν and k is an Ahlfors' set Ω_a of simple a_1-, a_2-, $\ldots$, a_q-points. Applying inequality (2.3.9') and inequality (2.3.5) of Theorem 2 to the points $z_k(a_\nu)$ we get with a constant K_5 depending only on a_ν, D_ν, D'_ν

$$|w'(z_k(a_\nu))| \geq \frac{K_5}{\varphi(r)} \frac{A^{1/2}}{r}.$$

Since $\varphi(r)$ is an arbitrary function we can rewrite the last inequality as (2.3.4). Theorem 1 is proved.

To prove Theorem 3 we construct domains D_ν involving Γ_ν and D'_ν similarly as in Theorem 2 and take as Γ_ν^k the simple islands over Γ_ν which lie in D_ν^k in the proof of Theorem 2. Since the number of such islands is a magnitude $n_0(r, D'_\nu) - o[A(r)]$ we have that the totality of these islands taken for all ν and k is an Ahlfors' set Ω_Γ of simple islands over $\Gamma_1, \Gamma_2, \ldots, \Gamma_q$. Applying inequality (2.3.9) and inequality (2.3.5) of Theorem 2 to these islands Γ_ν^k we get with a constant K_6 depending only on Γ_ν, D_ν, D'_ν

$$l(w^{-1}(\Gamma_\nu^k)) \le \frac{K_6\varphi(r)r}{A^{1/2}(r)}.$$

Since $\varphi(r)$ is an arbitrary function we can rewrite the last inequality as (2.3.5). Theorem 3 is proved.

Remark 3. The Ahlfors' Theory is properly adjusted for the study *regular exhausted covering surfaces*. In addition, the so-called *normal exhausted covering surfaces* were studied mainly in the Romanian school, see Stoilov [1], [2] and Andreian-Cazacu [1]. Since the present item studies the pre-images of the regular exhausted covering surfaces it seems now natural to ask: what can be said on the pre-images of normal exhausted covering surfaces? Particularly we expect that when the normal exhausted covering surfaces are simultaneously regular exhausted there must be interesting regularities related to the behavior of the pre-images resembling the known Littlewood's Property of entire functions.

2.4. Bounds of some integrals

2.4.1. We prove below some estimates which will be used in the successive sections.

Lemma 1. *Let* $w(z) \in M_R$, $r < r_1 < r_2 < r_3 < R$. *Then*

$$\int\int_{|z|<r} \frac{|w''(z)|}{|w'(z)|} d\sigma \le 32\pi r_2 \max\left(\ln \frac{r_2^2}{r_2^2 - r^2}; \frac{1}{\ln \frac{r_2}{r_1}}\right)[T(r_3, w)$$

$$+ O(\ln^+ T(r_3, w) + \ln^+ \frac{1}{r_3 - r_2} + \ln \frac{r_2}{r_1})], \quad r \to R.$$

$$(2.4.1)$$

Proof. By virtue of the Poisson–Jensen formula we get

$$
\iint_{|z|<r} \frac{|w''(z)|}{|w'(z)|} d\sigma \;\le\; \iint_{|z|<r} \left(\frac{1}{2\pi} \int_0^{2\pi} \left|\ln |w'(r_1 e^{i\vartheta})|\right| \frac{2r_1}{|r_1 e^{i\vartheta} - t e^{i\varphi}|^2} d\vartheta \right) d\sigma
$$

$$
+ \sum_{|a_m|\le r_1} \iint_{|z|<r} \left| \frac{1}{z-a_m} + \frac{1}{z - r_1^2/a_m} \right| d\sigma
$$

$$
+ \sum_{|b_n|\le r_1} \iint_{|z|<r} \left| \frac{1}{z-b_m} + \frac{1}{z - r_1^2/b_n} \right| d\sigma,
$$

where a_n and b_m are respectively zeros and poles of the function $w'(z)$.
Since

$$
\iint_{|z|<r} \frac{2r_1}{|r_1 e^{i\vartheta} - t e^{i\varphi}|^2} d\sigma \;=\; \int_0^r \left(\int_0^{2\pi} \frac{2r_1 \, d\varphi}{|r_1 e^{i\vartheta} - t e^{i\varphi}|^2} \right) t \, dt
$$

$$
= \int_0^r \frac{4\pi r_1 t}{r_1^2 - t^2} dt = 2\pi r_1 \ln \frac{r_1^2}{r_1^2 - r^2}
$$

and for arbitrary a

$$
\iint_{|z|<r} \frac{d\sigma}{|z-a|} \le 4\pi r
$$

we have

$$
\iint_{|z|<r} \frac{|w''(z)|}{|w'(z)|} d\sigma \;\le\; 2\pi r_1 \ln \frac{r_1^2}{r_1^2 - r^2} [(m(r_1,0,w') + m(r_1,\infty,w')]
$$

$$
+ 8\pi r[n(r_1,0,w') + n(r_1,\infty.w')]
$$

$$
\le\; 2\pi r_1 \ln \frac{r_1^2}{r_1^2 - r^2} [m(r_1,0,w') + m(r_1,\infty,w')]
$$

$$
+ \frac{8\pi r}{\ln \frac{r_2}{r_1}} [N(r_1,0,w') + N(r_1,\infty.w') + O(\ln \frac{r_2}{r_1})]
$$

$$
\le\; 8\pi r_2 \max \left(\ln \frac{r_2^2}{r_2^2 - r^2}; \frac{1}{\ln \frac{r_2}{r_1}} \right)
$$

$$
\times [T(r_2,w') + T(r_2,\frac{1}{w'})] + O(\ln \frac{r_2}{r_1})]
$$

$$
\le\; 16\pi r_2 \max \left(\ln \frac{r_2^2}{r_2^2 - r^2}; \frac{1}{\ln \frac{r_2}{r_1}} \right)
$$

$$
\times [T(r_2,w') + O(1 + \ln \frac{r_2}{r_1})]. \tag{2.4.2}
$$

Now taking into account the known connections between $T(r.w)$ and $T(r, w')$ we obtain (2.4.1).

Lemma 1'. *Let* $w(z) \in M$, $r < r_1 < r_2 < r_3 < R$. *Then there exist some constants* $K < \infty$ *and* $C \in (0,1)$ *such that for* $r \in E^*$

$$\int\int_{|z|<r} \frac{|w''(z)|}{|w'(z)|} d\sigma \leq KrA(r), \qquad (2.4.1')$$

holds, where E^* *is a set of lower logarithmic density* C, *i.e.*

$$\lim_{r\to\infty} \inf \frac{1}{r} \int_{E^*\cap(0,r)} d\ln t = C.$$

Remark. The magnitudes K and C and the set E^* are the same as in Miles' article [1].

The proof of this result can be found in the author's article [9]. The main point of its proof was given by Miles [1] which used the article of Hayman [3] which in turn used the long article of Borel [1]. We will not give this long proof in this small book! The reader is referred to the above mentioned articles.

Lemma 2. *Let* $w(z) \in M_R$, $c = \text{const} > 1$. *Then with a constant* $K(c) < \infty$ *depending only on* c

$$\int_0^{2\pi} \int_0^r \left| \frac{\partial}{\partial t} \arg \frac{\partial}{\partial t} w(te^{i\varphi}) \right| dt d\varphi = K(c)\widetilde{T}(cr, w), \quad r > r_0, \qquad (2.4.3)$$

holds, where

$$\widetilde{T}(r) = \int_0^r \frac{T(t)}{t} dt.$$

Proof. We denote the zeros and the poles of $w'(z)$ by a_m and b_n respectively. Also we denote $s_1 = (1 + (c-1)/4)r$, $s_2 = (1 + (c-1)/2)r$ and $s_3 = (1 + 3(c-1)/4)r$. Then using the Poisson–Jensen formula we get

$$\int_0^{2\pi} \int_0^r \left| \frac{\partial}{\partial t} \arg \frac{\partial}{\partial t} w(te^{i\varphi}) \right| dt d\varphi$$

$$= \int_0^{2\pi} \int_0^r \left| \frac{\partial}{\partial t} \arg w'(te^{i\varphi}) \right| dt d\varphi$$

$$\leq \int_0^r \int_0^{2\pi} \left(\frac{1}{2\pi} \int_0^{2\pi} |\ln |w'(s_1 e^{i\vartheta})|| \frac{2s_1}{|s_1 e^{i\vartheta} - te^{i\varphi}|} d\vartheta \right) d\varphi dt$$

$$+ \sum_{|a_m| \leq s_1} \int_0^{2\pi} \int_0^r \left| \frac{\partial}{\partial t} \arg \left(\frac{s_1^2 - \overline{a}_m z}{s_1(z - a_m)} \right) \right| dt d\varphi$$

$$+ \sum_{|b_n| \leq s_1} \int_0^{2\pi} \int_0^r \left| \frac{\partial}{\partial t} \arg \left(\frac{s_1^2 - \overline{b}_n z}{s_1(z - b_n)} \right) \right| dt d\varphi = J_1 + J_2 + J_3.$$

$$(2.4.4)$$

Obviously for $r > r_0$

$$J_1 \leq 2\pi \int_0^r \frac{2s_1}{s_1^2 - t^2} dt \, [m(s_1, 0, w') + m(s_1, \infty, w')]$$

$$\leq 4\pi \frac{s_1 r}{s_1^2 - r^2} [2T(s_1, w') + O(1)] \leq K(c) T(s_1, w'). \quad (2.4.5)$$

As the function $(s_1^2 - \overline{a}_m z)/(s_1(z - a_m))$ maps $\{z : \arg z = \text{const}\}$ to a straight line or to a circle, by the geometric meaning of the integrand of J_2 we find

$$\int_0^r \left| \frac{\partial}{\partial t} \arg \left(\frac{s_1^2 - \overline{a}_m z}{s_1(z - a_m)} \right) \right| dt \leq 2\pi.$$

Hence

$$J_2 \leq 4\pi n(s_1, 0, w') \leq \frac{4\pi}{\ln(s_2/s_1)} N(s_2, 0, w') = K(c) N(s_2, 0, w'), \quad (2.4.6)$$

and similarly

$$J_3 \leq 4\pi n(s_1, 0, w') \leq K(c) N(s_2, \infty, w'). \quad (2.4.6')$$

Using the known inequality

$$T(s_2, w') \leq 2T(s_2, w) + m\left(s_2, \frac{w'}{w} \right) \quad (2.4.7)$$

and the Lemma on Logarithmic Derivative we have

$$
\begin{aligned}
T(s_2, w') &\leq K(c)T(s_3, w) \\
&\leq \frac{K(c)}{\ln(cr/s_3)}\widetilde{T}(cr, w) = K(c)\widetilde{T}(cr, w), \quad r > r_0,
\end{aligned}
$$

so that Lemma 2 follows now from (2.4.4) to (2.4.6) and (2.4.6′).

Repeating the proof of Lemma 2 almost word by word we get

Lemma 2′. *Let $w(z) \in M_R$, $c = \mathrm{const} > 1$. Then with a constant $K(c) < \infty$ depending only on c*

$$
\int_0^{2\pi} \left| \frac{\partial}{\partial\varphi} \arg w'(te^{i\varphi}) \right| d\varphi \leq K(c)T(cr, w), \quad r > r_0 \tag{2.4.8}
$$

holds.

Lemma 3. *Let $w(z) \in M_R$, $c = \mathrm{const} > 1$. Then with a constant $K(c) < \infty$ depending only on c*

$$
\int_0^{2\pi} \int_0^r \left| \frac{\partial}{\partial\varphi} \arg \frac{\partial}{\partial\varphi} w(re^{i\varphi}) \right| dtd\varphi = K(c)\widetilde{T}(cr, w), \quad r > r_0 \tag{2.4.9}
$$

holds.

Proof. We have

$$
\int_1^r \int_0^{2\pi} \left| \frac{\partial}{\partial\varphi} \arg \frac{\partial}{\partial\varphi} w(te^{i\varphi}) \right| d\varphi \frac{dt}{t} \leq \int_1^r \int_0^{2\pi} \left| \frac{\partial}{\partial\varphi} \arg w'(te^{i\varphi}) \right| d\varphi \frac{dt}{t} + 2\pi \ln r
$$

so that applying Lemma 2′ to the last integral and using the Lemma on the Logarithmic Derivative we deduce that

$$
\begin{aligned}
&\int_1^r \int_0^{2\pi} \left| \frac{\partial}{\partial\varphi} \arg \frac{\partial}{\partial\varphi} w(te^{i\varphi}) \right| d\varphi \frac{dt}{t} \\
&\leq K(c) \int_1^r T\left(\left(1 + \frac{c-1}{4}\right) t, w' \right) t^{-1} dt + 2\pi \ln r \\
&\leq K(c) \int_1^r \left[2T\left(\left(\frac{c-1}{4} + 1\right) t, w \right) + m\left(\left(\frac{c-1}{4} + 1\right) t, \frac{w'}{w} \right) \right] \frac{dt}{t} \\
&\quad + 2\pi \ln \leq K(c) \int_1^r T\left(\left(1 + \frac{c-1}{2}\right) t, w \right) \frac{dt}{t} \\
&\quad + 2\pi \ln r \leq K(c)\widetilde{T}(cr, w).
\end{aligned}
$$

2.4.2. We prove below some estimates concerning the spherical length

$$L(r) := \int_0^{2\pi} \frac{|w'(z)|}{1 + |w(z)|^2} r d\varphi.$$

Lemma 4. *Let* $w(z) \in M$. *Then*

$$\int_1^r \int_0^{2\pi} \frac{|w'(z)|}{1 + |w(z)|^2} dt d\varphi = o\left(\tilde{T}(cr, w)\right). \tag{2.4.10}$$

Equation (2.4.10) holds because the integral is less than or equal

$$\left(\int_1^r \int_0^{2\pi} \frac{dt}{t} d\varphi\right)^{1/2} A^{1/2}(r, w) \leq (2\pi \ln r)^{1/2} \left(\frac{T(s_1, w)}{\ln(s_1/r)}\right)^{1/2}$$

$$= o\left(\tilde{T}(cr, w)\right).$$

Lemma 5. *Let* $w(z) \in M$, $c = const > 1$. *Then*

$$\int_0^r L(t, w) dt \leq \pi r A^{1/2}(r), \quad r > 0, \tag{2.4.11}$$

$$\int_0^r L(t, w) dt = O[r T^{1/2}(r)], \quad r \to \infty, \tag{2.4.12}$$

$$\int_0^r L(t, w') dt = O[r T^{1/2}(r)], \quad r \to \infty, \tag{2.4.13}$$

Indeed, using the Cauchy–Buniacovski inequality we have

$$\int_0^r L(t, w) dt = \int_1^r \int_0^{2\pi} \frac{|w'(z)|}{1 + |w(z)|^2} t dt d\varphi$$

$$\leq \left(\int_1^r \int_0^{2\pi} t dt d\varphi\right)^{1/2} \left(\int_1^r \int_0^{2\pi} \frac{|w'(z)|^2}{(1 + |w(z)|^2)^2} t dt d\varphi\right)^{1/2}$$

$$= \pi r A^{1/2}(r).$$

Taking into account

$$T^0(cr, w) \geq \int_r^{cr} \frac{A(t)}{t} dt \geq A(r) \ln c$$

and

$$T^0(cr, w) = T(cr, w) + O(1), \quad r \to \infty$$

we get (2.4.12).

Similarly, taking instead of c the constant $1 + (c-1)/2$ we get

$$\int_0^r L(t, w') dt$$

$$\leq \pi r \ln^{-1} \left(1 + \frac{c-1}{2} \right) \left(rT \left(\left(1 + \frac{c-1}{2} \right) r, w' \right) + O(1) \right)^{1/2},$$

$$r \to \infty,$$

and applying (2.4.7) and the Lemma on the Logarithmic Derivative we get (2.4.13).

2.4.3. We need also the following useful lemma.

Lemma 6 (see for instance Petrenko [1], Lemma 1.3.1). *Suppose that $w(z) \in M$ has bounded lower order λ. Then for every $c = $ const > 1 there exists a sequence $r_n = r_n(c) \to \infty$ such that for $r_n > r_0$*

$$T(cr_n, w) \leq c^{\lambda+1} T(r_n, w) \tag{2.4.14}$$

holds.

Proof. We suppose that for a given constant $c > 1$ (2.4.14) does not hold, i.e. starting with an $r > r_0$

$$T(cr, w) > c^{\lambda+1} T(r_n, w)$$

holds. Then taking $r = r_n = r_0 c^n$ we obtain

$$\begin{aligned} T(r_n, w) &= T(c^n r_0, w) \geq c^{\lambda+1} T(c^{n-1} r_0, w) \\ &\geq c^{2(\lambda+1)} T(c^{n-2} r_0, w) \geq \cdots \geq c^{n(\lambda+1)} T(r_0, w) \\ &= \frac{r_n^{\lambda+1}}{r_0^{\lambda+1}} T(r_0, w) = \text{const } r_n^{\lambda+1}. \end{aligned}$$

Since the last inequality implies

$$1 + \lambda \leq \lim_{r \to \infty} \inf \frac{\ln T(r, w)}{\ln r} := \lambda.$$

we come to a contradiction which proves (2.4.14).

CHAPTER 3

Γ-LINES' APPROACH IN THE THEORY OF MEROMORPHIC FUNCTIONS

Introduction. Nevanlinna's Value Distribution Theory and Proximity Property of a-points study, respectively, the numbers and mutual locations of a-points of arbitrary meromorphic functions in $\mathbb{C}$. The main conclusion of the Theory and the Property is true also for meromorphic functions with "fast growth" in the unit disk, but not for functions with "slow growth" in general. The last circumstance is an essential gap in Complex Analysis since many known classes of functions do have "slow growth", (among them classes of Bounded Functions, H^p, Dirichlet, Blaschke Products etc.). In this chapter, we offer a novel, geometric approach to the study of Value Distribution and mutual locations of a-points for arbitrary meromorphic functions including those with slow growth. Our approach is based on the above results related to Γ-lines.

3.1. Principle of Closeness of Sufficiently Large Sets of a-Points of Meromorphic Functions

The classical Value Distribution Theory of Nevanlinna [1] and the Theory of Covering Surfaces of Ahlfors [3] describe the numbers of the a-points of functions $w(z)$ meromorphic in $\mathbb{C}$. The main Deficiency Relations in these theories qualitatively asserts that for any such function $w(z)$ there is a set $G \subset \mathbb{C}$ of "good", i.e. non-deficient, values of $w(z)$ such that G "almost coincides" with $\mathbb{C}$ and for any value $a \in G$ the counting functions $n(r, a)$ is close to the characteristic $A(r)$. In other words, the majority of complex values occur in G and for a and $b \in G$ the magnitudes $n(r, a)$ and $n(r, b)$ are asymptotically close. The last phenomenon we will refer as *the Nevanlinna Property*.

As to functions $w(z)$ meromorphic in the unit disk, the Nevanlinna Property is true also for functions with *fast growth*, i.e. when

$$\limsup_{r \to 1} A(r)(1 - r) = \infty, \qquad (3.1.1)$$

where

$$A(r) = \frac{1}{\pi} \int \int_{|z| < r} \frac{|w'(z)|^2}{(1 + |w(z)|^2)^2} d\sigma$$

is the spherical characteristic function.

For the same classes of functions it is possible to describe mutual arrange-
ment (m.a.) of *a*-points for different values *a* (see "Proximity Property of
a-points" in Chapter 4 and also in the author's papers [6, 9, 12, 13, 15, 16]).
In addition to the Nevanlinna Property, the Proximity Property states that
not only the numbers but also the geometric locations of these *a* and *b*-points
are close as well.

There is a certain gap in Function Theory in the case when $A(r)$ has *slow
growth*, i.e. when

$$A(r) = O\left(\frac{1}{1-r}\right), \quad r \to 1. \tag{3.1.2}$$

In fact, very few results have been obtained not only for m.a., but also
for the number of *a*-points. For the well-known cases (for classes of bounded
functions, H^p, Dirichlet and so on) one can only estimate the Blaschke sums
for the zeros and cannot even compare the number of *a*-points for different
$a \in \mathbb{C}$. Indeed, these numbers can be strikingly different; for instance, the
functions with slow growth can omit taking values $a \in E(a)$ for "large" sets
$E(a) \subset \mathbb{C}$.[8] Therefore the Nevanlinna Deficiency Relation is no more true
for functions with slow growth and we have no Nevanlinna Property and no
Proximity Property in the general case.

Below we offer a novel, geometric approach to the study of Value Distribution and Proximity Property for all meromorphic functions in the complex
plane or in the unit disk, including those with slow growth.

The obtained results show different corollaries depending on the classes of
functions. If we consider, say, the classical case, that is functions in the complex plane or functions in the unit disk with fast growth, then the corollaries
coincide qualitatively with the Nevanlinna Property and the Proximity Property. If we consider functions in the unit disk with slow growth, the corollaries
show some, say, weakened forms of the Properties. An interesting point is
that while considering particular classes of functions with slow growth, such
as known Hardy and Dirichlet classes, Blaschke products, we describe the
weakened Properties in terms of the characteristics of these classes.

---

[8]On the contrary, for functions with fast growth the set $E(a)$ may omit at most two
points, according to the Deficiency Relation. Thus, we observe a striking distinction in
behavior of the numbers of *a*-points for functions with slow growth and those with fast
growth.

The main idea here is very simple and based on the study of Γ-lines. Namely, if $a, b \in \Gamma$, where Γ is a complex curve, then by estimating the length of the Γ-lines we obtain estimates of distances between a- and b-points.

It is interesting that this approach permits to describe the numbers and m.a. using a very simple background accessible even for those who have just started to learn Complex Analysis.

Another principal conclusion is that we immediately obtain a kind of Second Fundamental Theorem of Nevanlinna for any class of functions in the unit disk, if we have the above estimate for the lengths of Γ-lines for the same class of functions.

In Section 3.3, we prove some results resembling Nevanlinna's Second Fundamental Theorem. These results evaluate $\sum \mathcal{M}_i(r)$, where the magnitudes $\mathcal{M}_i(r)$ reflects the growth of functions in different directions.

3.1.1. Characteristic functions describing the numbers and the m.a. of a-points.

Let $w(z)$ be a meromorphic function in $|z| < R^* \le \infty$ and let $D(r) = \{z : |z| < r\}. r < R^*$. Further. let $\{a_1, b_1, a_2, b_2, \ldots, a_n, b_n\}$ be a collection of pairwise different complex values. and let A be the set of all w-points, $w \in \{a_1. b_1. a_2. b_2. \ldots. a_n. b_n\}$ with counting multiplicity. As counting functions of the set $A \cap D(r)$ for the function $w(z)$ we shall consider

$$\mathcal{N}_A(r) = \sum_{\nu=1}^{n} (\mathcal{N}(r, a_\nu) + \mathcal{N}(r. b_\nu)), \tag{3.1.3}$$

where

$$\mathcal{N}(r, w) = \int_0^r n(t, w) \, dt$$

and $n(r, w)$ is the usual number of w-points in the disk $|z| < r$ with counting multiplicity.

Remark. In the case of functions given in the complex plane one can easily estimate the sum in (3.1.3) in terms of Nevanlinna's characteristic function $T(r)$, by the use of Nevanlinna's theory. In the case of functions given in the unit disk it is not possible because of the reason mentioned above.

Now let Γ be a smooth Jordan curve in $\mathbb{C}$ passing through given points $a_1, b_1, a_2, b_2, \ldots, a_n, b_n$ ($n < \infty$) in this order. We denote by $z_i(w)$ the w-points of the function $w(z)$; by $A = A(\Gamma, r)$ the set of all $z_i(w)$ in $|z| < r$ with counting multiplicities, where $w \in \{a_1, b_1, a_2. b_2, \ldots, a_n, b_n\}$.

First we shall make the following.

Assumption. $w^{-1}(\Gamma)$ does not pass through any points $z \in D(r)$, $z \notin A$ at which $w'(z) = 0$.

We denote by Γ_{a_ν, b_ν} the part of the curve Γ lying between the points a_ν and b_ν, $\nu = 1, 2, \ldots, n$. Then the set $w^{-1}(\Gamma_{a_\nu, b_\nu})$ lying in $D(r)$ consists of some curves of the types L_1, L_2, L_3 which, due to the Assumption are completely determined as follows:

(A) Every curve of type L_1 connects an a_ν-point with b_ν-point (we shall prescribe to these connected a_ν- and b_ν-points the same index j). Hence $|z_j(a_\nu) - z_j(b_\nu)|$ is less than or equal to the length of this curve.

(B) Every curve of type L_2 connects a point $z_i(a_\nu)$ (or $z_i(b_\nu)$) with a point on the boundary $D(r)$. Hence the magnitude $r - |z_i(a_\nu)|$ (or $r - |z_i(b_\nu)|$) is less than or equal to the length of this curve.

(C) Every curve of type L_3 has no endpoint coinciding with a $z_i(a_\nu)$ or a $z_i(b_\nu)$.

We recall that every a_ν-point or b_ν-point above is numerated counting multiplicities.

Let $\tilde{A} = \tilde{A}(\Gamma, r)$ be the collection of all pairs $(z_j(a_\nu), z_j(b_\nu))$ from (A) for all ν and let

$$C(r, \tilde{A}) = \sum_{(z_j(a_\nu), z_j(b_\nu)) \in \tilde{A}} |z_j(a_\nu) - z_j(b_\nu)|. \tag{3.1.4}$$

We shall call $C(r, \tilde{A})$ the closeness function of a- and b-points.

Note that only one of the two points $z_i(a_\nu)$ and $z_i(b_\nu)$ can be an endpoint of the given curve of type L_2. Let $A^* = A^*(\Gamma, r)$ be the set of all such *lonely* points $z_i(a_\nu)$ or $z_i(b_\nu)$ from (B) and let

$$\mathcal{N}_{A^*}(r) = \sum_{\nu=1}^{n} (\mathcal{N}^*(r, a_\nu) + \mathcal{N}^*(r, b_\nu)), \tag{3.1.5}$$

where

$$\mathcal{N}^*(r, w) = \int_0^r n^*(t, w) dt$$

and $n^*(r, w)$ is the number of w-points from (B) with counting multiplicities. We call $\mathcal{N}_{A^*}$ the counting function of lonely a- and b-points.

For the close a-and b-points from $\tilde{A}$ we shall consider also

$$\mathcal{N}_{\tilde{A}}(r) = \sum_{\nu=1}^{n}(\tilde{\mathcal{N}}(r, a_\nu) + \tilde{\mathcal{N}}(r, b_\nu)), \qquad (3.1.5')$$

where

$$\tilde{\mathcal{N}}(r, w) = \int_0^r \tilde{n}(t, w)dt$$

and $\tilde{n}(r, w)$ is the number of w-points from $\tilde{A}$ with counting multiplicities. We call $\mathcal{N}_{\tilde{A}}$ the counting function of close a- and b-points.

3.1.2. Connections between the closeness function and the function of lonely points with the lengths of Γ-lines. It is obvious that

$$\mathcal{N}_A(r) = \sum_{z_j(a)\in A} (r - |z_j(a)|), \quad \mathcal{N}_{A^*}(r) = \sum_{z_j(a)\in A^*} (r - |z_j(a)|),$$

$$\mathcal{N}_{\tilde{A}}(r) = \sum_{z_j(a)\in\tilde{A}} (r - |z_j(a)|) \qquad (3.1.6)$$

and that $|z_j(a_\nu) - z_j(b_\nu)|$, $r - |z_j(a_\nu)|$ and $r - |z_j(a_\nu)|$ are less than or equal to the above mentioned lengths of the curves. Therefore denoting by $L(r, \Gamma)$ the total Euclidean length of the curves $w^{-1}(\Gamma)$ in $D(r)$ we come to the following.

Fundamental inequality. *For any function $w(z)$ meromorphic in $|z| < R^* \leq \infty$ and any curve Γ the set A of all a_ν- and b_ν-points can be represented as $\tilde{A} \cup A^*$ so that for the closeness function and the function of lonely a- and b-points we have*

$$\mathcal{C}(r, \tilde{A}) + \mathcal{N}_{A^*}(r) \leq L(r, \Gamma), \quad r < R^*. \qquad (3.1.7)$$

The above definitions (A), (B) and (C), and consequently (3.1.7), are true also for the case where the above assumption does not hold. In this case, the separation of the close pairs and lonely points can be accomplished by different ways, we will choose any.

3.1.3. Connections between the Fundamental inequality and Nevanlinna's Value Distribution Theory.

According to Nevanlinna's Second Fundamental Theorem, for meromorphic functions in $\mathbb{C}$ and given pairwise different values $a_1, a_2, \ldots, a_n$

$$\sum_{\nu=1}^{n}[T(r) - N(r, a_\nu)] \leq 2T(r) + o(T(r)), \quad r \to \infty, \quad r \notin E$$

is valid, where E is a set of finite measure. Due to this estimate one obtains the main conclusion of Nevanlinna's Theory on the closeness of $N(r, a_\nu)$ to $T(r)$ for the majority of the values a_ν.

This assertion almost coincides with the following one: for a given set of pairwise different values $a_1, b_1, a_2, b_2, \ldots, a_n, b_n$ the inequality

$$\sum_{\nu=1}^{n} |N(r, b_\nu) - N(r, a_\nu)| \leq 2T(r) + o(T(r)), \quad r \to \infty, \quad r \notin E$$

holds.

Let us consider now the expression

$$\sum_{\nu=1}^{n} |\mathcal{N}(r, b_\nu) - \mathcal{N}(r, a_\nu)|,$$

which we call the *Nevanlinna Difference*.

Obviously, if $a_1, b_1, a_2, b_2, \ldots, a_n, b_n$ occur on a curve Γ in the same order, then we have

$$|\mathcal{N}(r, b_\nu) - \mathcal{N}(r, a_\nu)| = |\mathcal{N}^*(r, b_\nu) - \mathcal{N}^*(r, a_\nu)| \leq \mathcal{N}^*(r, b_\nu) + \mathcal{N}^*(r, a_\nu),$$

so that the Fundamental Inequality implies

$$\begin{aligned} \sum_{\nu=1}^{n} |\mathcal{N}(r, b_\nu) - \mathcal{N}(r, a_\nu)| &\leq \sum_{\nu=1}^{n}(\mathcal{N}^*(r, b_\nu) + \mathcal{N}^*(r, a_\nu)) \\ &= \mathcal{N}_{A^*}(r) \leq L(r, \Gamma). \end{aligned} \quad (3.1.7')$$

Thus, inequality (3.1.7′) offers an essentially different approach for evaluating the Nevanlinna Difference for any functions in the complex plane or in the unit disk. Further, for evaluating these differences we just need to estimate the length of the corresponding Γ-lines. Following this idea, we give the upper bounds for $L(r, \Gamma)$ and come qualitatively to the above Nevanlinna's conclusions in the case of functions in the complex plane. However, we obtain simultaneously a description of m.a. of a-points, that is the Proximity Property.

In Section 3.2, we give estimates of this difference for various known classes of functions and consequently for the considered classes we obtain some assertions analogous to the Second Fundamental Theorem of Nevanlinna.

3.1.4. Sufficiently large sets: the principle of closeness of sufficiently large sets of a-points

Definition 1. Let the function w be meromorphic in $|z| < R^* \leq \infty$ and let $\varepsilon > 0$. Consider the above collection $\{a_1, b_1, a_2, b_2, \ldots, a_n, b_n\}$ belonging to the curve Γ. For a given r, $0 < r < R^*$, we call the set $A = A(\Gamma, r)$ *sufficiently large* if

$$\mathcal{N}_A(r) \geq \frac{L(r, \Gamma)}{\varepsilon}. \tag{3.1.8}$$

The principle in the title is reflected in the following.

Theorem 1 (see Barsegian [17]) *Let w be a function meromorphic in $z : |z| < R^*$, $R^* \leq \infty$, Γ be a smooth Jordan curve passing through given points $a_1, b_1, a_2, b_2, \ldots, a_n, b_n$ $(n < \infty)$ in this order, $\varepsilon = const > 0$. If the set $A(\Gamma, r) = \tilde{A}(\Gamma, r) \cup A^*(\Gamma, r)$ is sufficiently large for a given $r < R^*$ then*

$$\frac{\mathcal{N}_{A^*}(r)}{\mathcal{N}_A(r)} \leq \varepsilon, \tag{3.1.9}$$

and

$$\frac{\mathcal{C}(r, \tilde{A})}{\mathcal{N}_{\tilde{A}}(r)} \leq \varepsilon. \tag{3.1.10}$$

Remark 1. If ε is small enough, then by (3.1.9) the majority of the a- and b-points belong to $\tilde{A}$.

Remark 2. Using inequality (3.1.7′) we can change inequality (3.1.9) by the following inequality, more usual in Nevanlinna's theory,

$$\frac{\sum_{\nu=1}^{n} |\mathcal{N}(r, b_\nu) - \mathcal{N}(r, a_\nu)|}{\sum_{\nu=1}^{n} |\mathcal{N}(r, b_\nu) + \mathcal{N}(r, a_\nu)|} \leq \varepsilon, \tag{3.1.9′}$$

which clearly shows that the magnitudes $\mathcal{N}(r, b_\nu)$ and $\mathcal{N}(r, a_\nu)$ are comparatively close.

Theorem 1 has a simple meaning. By Remark 1 the majority of the a- and b-points belong to $\tilde{A}$, and by (3.1.10) a- and b-points from $\tilde{A}$ are close to one other (namely, the closeness function of the close points is essentially less than the counting function of the same close points).

Obviously, for any meromorphic function in $|z| < R^*$, for any $\varepsilon > 0$ there exists a curve Γ such that for each given $r < R^*$ we can choose a certain collection $\{a_1, b_1, a_2, b_2, \ldots, a_n, b_n\} \subset \Gamma$ for which the set $A = A(\Gamma, r)$ is sufficiently large. Also we note that as such a Γ in this sufficiently large set we can choose, for instance, a curve passing through an arbitrary point belonging to $w(D(r))$. With the same Γ by adding new pairs a_ν, b_ν we can get sufficiently large sets $A(\Gamma, r')$ for any $r' > r$.

In the definition of the sufficiently large sets we use the length $L(r, \Gamma)$. However we can define sufficiently large sets making use of the above bounds for the length. This will allow us to describe the numbers and m.a. of a-points for those classes of functions for which we have similar bounds. Sometimes the terms of above bounds can appear to be more convenient.

For arbitrary meromorphic functions according to the "Tangent Variation Principle" (see Section 1.2) we have

$$L(r, \Gamma) \le 3(v(\Gamma) + 1)(B(r) + 2\pi r).$$

Therefore if for given collection $\{a_1, b_1, a_2, b_2, \ldots, a_n, b_n\} \subset \Gamma$

$$\mathcal{N}_A(r) \ge \frac{3(v(\Gamma) + 1)(B(r) + 2\pi r)}{\varepsilon} \tag{3.1.11}$$

holds then the set $A(\Gamma, r) = \tilde{A}(\Gamma, r) \cup A^*(\Gamma, r)$ is sufficiently large, so that applying Theorem 1 we get the following.

Theorem 2. *Let the conditions of Theorem 1 be satisfied and additionally inequality* (3.1.11). *Then inequalities* (3.1.9) *and* (3.1.10) *hold.*

For functions with "slow" growth, satisfying (3.1.2), the following estimate is valid.

Theorem 3. *Let $w(z)$ be a function meromorphic in $D(1)$ and satisfying the relation* (2). *Then*

$$L(r, \Gamma) \le c\left(\frac{1}{1-r}\log\frac{1}{1-r}\right) + c', \quad r < 1, \tag{3.1.12}$$

where c and c' are some constants depending only on w and Γ.

As above we come to the following.

Theorem 4. *Let the conditions of Theorem 1 be satisfied and additionally*

$$\mathcal{N}_A(r) \geq \frac{1}{\varepsilon}\left[c\left(\frac{1}{1-r}\log\frac{1}{1-r}\right)+c'\right]$$

be valid. Then inequalities (3.1.9) and (3.1.10) hold.

The Γ-lines of conformal mappings in the unit disk were studied in numerous articles (see Section 1.2). Finally Öyma [1] has obtained the simple proof of the following.

Theorem A. *Let a function $w(z) = z + \alpha_2 z^2 + \cdots$ be holomorphic and univalent in the unit disk D and let Γ be a smooth Jordan curve. Then the length $L(1,\Gamma)$ of the pre-image $w^{-1}(\Gamma)$ satisfies*

$$L(1,\Gamma) < c_1. \tag{3.1.13}$$

From Theorem A arguing as above we obtain the following.

Theorem 5. *Let the conditions of Theorem 1 be satisfied and additionally*

$$\mathcal{N}_A(r) \geq \frac{c_1}{\varepsilon}.$$

Then inequalities (3.1.9) and (3.1.10) hold.

3.1.5. Proofs. To prove Theorem 1 it is enough to observe that due to the above definition of "sufficiently large" sets the following inequality is valid

$$\mathcal{C}(r,\tilde{A}) + \mathcal{N}_{A^\bullet}(r) \leq \varepsilon\mathcal{N}_A(r), \tag{3.1.14}$$

which proves (3.1.9). Now (3.1.10) follows from (3.1.14) if we replace $\mathcal{N}_A(r)$ by $\mathcal{N}_{\tilde{A}}(r)+\mathcal{N}_{A^\bullet}(r)$ and take the following inequality into account: for $a,b,c > 0$ and $a \leq b$

$$\frac{a+c}{b+c} \geq \frac{a}{b}.$$

Theorems 2 and 5 are proved similarly. Theorem 4 also is proved similarly, taking Theorem 3 into account. Now we prove Theorem 3.

Let $r' = (1 + 2r)/3$. Due to the inequality (2.3.1) of Section 2.3

$$B(r) = O\left(\frac{1}{1-r} T(r', w')\right), \quad r \to 1. \tag{3.1.15}$$

By a well-known inequality,

$$T(r', w') \leq 2T(r', w) + m(r', w'/w).$$

If $r_1' = (2 + r)/3$, then by the Lemma on the Logarithmic Derivative (see Section 2.1)

$$T(r', w') = 2T(r, w) + O\left[\log T(r_1', w) + \log \frac{1}{r_1' - r'}\right], \quad r \to 1.$$

Therefore, by (3.1.15)

$$B(r) = O\left(\frac{1}{1-r}\left[T(r_1', w) + \log \frac{1}{1-r}\right]\right), \quad r \to 1. \tag{3.1.16}$$

By Theorem 6.2 in Hayman's book [2], inequality (3.1.2) implies that

$$T(r, w) = O\left(\log \frac{1}{1-r}\right), \quad r \to 1,$$

Theorem 3 now follows from (3.1.16).

3.2. Integrated Version of the Principle: Connections with Known Classes of Functions – Qualitative and Quantitative Conclusions

3.2.1. The fundamental inequality (integrated version). Theorem 1 of the previous section gives a description of closeness of a-points for arbitrary functions $w(z)$ meromorphic in the unit disk. However, the terminology we use can appear to be alien for a given class of functions. Below we offer another integrated version of the principle of closeness of a-points which permits to describe the closeness in terms of characteristics of some known classes of functions.

Suppose that γ_ν, $\nu = 1, 2, \ldots, n$ and τ_ν, $\nu = 1, 2, \ldots, n$, are two collections of such sets that:

(1) for any $\nu = 1, 2, \ldots, n$, the sets γ_ν and τ_ν consist of continuous curves;

(2) for any $\nu = 1, 2, \ldots, n$ the set γ_ν intersects every circle $\Gamma(R) = \{w : |w| = R\}$ only in one point (which we denote $a_\nu(R)$) and also the set τ_ν intersects every circle $\Gamma(R)$ only in one point (which we denote $b_\nu(R)$);

(3) for any R the points $a_\nu(R)$ and $b_\nu(R)$ on $\Gamma(R)$ are arranged by increasing argument in the order $a_1(R)$, $b_1(R)$, $a_2(R)$, $b_2(R)$, \ldots, $a_n(R)$, $b_n(R)$, $a_1(R)$.

Applying now the fundamental inequality (3.1.7) of Section 3.1 for $w(z)$, $\Gamma = \Gamma(R)$ and $a_\nu = a_\nu(R)$, $b_\nu = b_\nu(R)$, $\nu = 1, 2, \ldots, n$ we obtain

$$\mathcal{C}(r, \tilde{A}(R)) + \mathcal{N}_{A^*(R)}(r)) \leq L(r. \Gamma(R)), \qquad (3.2.1)$$

where $A(R) := A(\Gamma(R), r)$ is the collection of all $a_\nu(R)$- and $b_\nu(R)$-points, respectively, $\tilde{A}(R) := \tilde{A}(\Gamma(R), r)$ and $A^*(R) := A^*(\Gamma(R), r)$.

Now observe that for any positive function $\psi(R)$ defined in $(0, \infty)$ the functions

$$\int_0^\infty \mathcal{N}_{A(R)}(r) \frac{dR}{\psi(R)}, \quad \int_0^\infty \mathcal{N}_{A^*(R)}(r) \frac{dR}{\psi(R)}, \quad \int_0^\infty \mathcal{N}_{\tilde{A}(R)}(r) \frac{dR}{\psi(R)}$$

correspondingly characterize the integral capacity of all $a_\nu(R)$, $b_\nu(R)$-points, of all lonely $a_\nu(R)$, $b_\nu(R)$-points, of all close $a_\nu(R)$. $b_\nu(R)$-points on the curves γ_ν and τ_ν. Further, the function

$$\int_0^\infty \mathcal{C}(r, \tilde{A}(R)) \frac{dR.}{\psi(R)}$$

characterizes the integral closeness of all $a_\nu(R)$. $b_\nu(R)$-points of $w(z)$ on these curves.

In this notation we come to the following.

Integrated fundamental inequality

$$\int_0^\infty \mathcal{C}(r, \tilde{A}(R)) \frac{dR.}{\psi(R)} + \int_0^\infty \mathcal{N}_{A^*(R)}(r) \frac{dR}{\psi(R)} \leq \int\int_{D(r)} \frac{|w'(z)|}{\psi(|w(z)|)} d\sigma, \quad (3.2.2)$$

which immediately follows from (3.2.1) and the main identity from Section 1.1:

$$\int_0^\infty L(r, \Gamma(R)) \frac{dR}{\psi(R)} = \int\int_{D(r)} \frac{|w'(z)|}{\psi(|w(z)|)} d\sigma. \qquad (3.2.3)$$

Arguing similarly to the proof of Theorem 1 in Section 3.1 we can introduce a concept analogous to "sufficiently large sets" and obtain some results similar to Theorem 1. We omit this.

3.2.2. Some qualitative conclusions from the integrated fundamental inequality for functions $w(z)$ meromorphic in $\mathbb{C}$. We shall show that the main conclusions of Nevanlinna Value Distribution Theory and the Proximity Property of a-points of functions meromorphic in $\mathbb{C}$ can be qualitatively derived from inequality (3.2.2) and the First Fundamental Theorem of Ahlfors' theory of Covering Surfaces. Using the inequality (3.2.2) with $\psi(R) = 1 + R^2$ we get

$$\int_0^\infty \mathcal{C}(r, \tilde{A}(R)) \frac{dR,}{1 + R^2} + \int_0^\infty \mathcal{N}_{A^*(R)}(r) \frac{dR}{1 + R^2}$$

$$\leq \int \int_{D(r)} \frac{|w'(z)|}{1 + |w(z)|^2} d\sigma \leq \pi r A^{1/2}(r). \qquad (3.2.4)$$

For simplicity suppose that our sets γ_ν and τ_ν are rays with the origin at the point $w = 0$. Then due to this theorem of Ahlfors we have

$$\int_0^\infty [n(r, a_\nu(R)) + n(r, b_\nu(R))] \frac{dR}{1 + R^2} = \pi A(r) + O(L(r)), \quad r \to \infty,$$

(for the definition of $L(r)$ see Section 2.1) and consequently

$$\int_0^\infty \mathcal{N}_{A(R)}(r) \frac{dR}{1 + R^2}$$

$$= \int_0^r \left\{ \sum_{\nu=1}^n \int_0^\infty [n(r, a_\nu(R)) + n(r, b_\nu(R))] \frac{dR}{1 + R^2} \right\}$$

$$= \pi n r A(r) + O\left(\int_0^r L(t) dt \right), \quad r \to \infty. \qquad (3.2.5)$$

This evidently shows that our lonely $a_\nu(R)$- and $b_\nu(R)$-points are taken very rarely in the general case, since by (3.2.4)

$$\int_0^\infty \frac{\mathcal{N}_{A^*(R)}(r)}{1 + R^2} dR$$

$$= \int_0^r \left\{ \sum_{\nu=1}^n \int_0^\infty [n^*(r, a_\nu(R)) + n^*(r, b_\nu(R))] \frac{dR}{1 + R^2} \right\} dr$$

$$\leq \pi r A^{1/2}(r). \qquad (3.2.6)$$

Now, since the numbers of lonely points are "small" we obtain that the majority of the $a_\nu(R)$- and $b_\nu(R)$-points occur in the set $\tilde{A}(R)$ in $D(r)$ together. Therefore the numbers of these $a_\nu(R)$- and $b_\nu(R)$-points are close to one another. This is quite similar to the main conclusion of Nevanlinna's and Ahlfors's theories. On the other hand, $a_\nu(R)$- and $b_\nu(R)$-points must also be geometrically close to one another according to the same inequality (3.2.4). The average distance $|z_i(a_\nu(R) - z_i(b_\nu(R)))|$ is a magnitude $O\left(rA^{-1/2}(r)\right)$ since the distance is defined by the ratio of the magnitudes

$$\int_0^\infty \mathcal{C}(r, \tilde{A}(R)) \frac{dR}{1 + R^2}, \quad \text{and} \quad \int_0^\infty [n(r, a_\nu(R)) + n(r, b_\nu(R))] \frac{dR}{1 + R^2}.$$

The last conclusion reflects the one mentioned in the introduction, "Proximity Property of a-points" of meromorphic functions in $\mathbb{C}$.

3.2.3. Analogs of the Second Fundamental Theorem of Nevanlinna and of the Proximity Property which already are valid for any function meromorphic in the complex plane and in the unit disk.
According to (3.1.7′) in Section 3.1 we can rewrite inequality (3.2.2) in the following form.

Theorem 1. *For arbitrary function meromorphic in $|z| < R^*$, $R^* \leq \infty$, for arbitrary continuous positive functions $\psi(R)$, and for the points $a_\nu(R)$ and $b_\nu(R)$, $\nu = 1, 2, \ldots, n$ defined in Section 3.2.1, the following inequality*

$$\int_0^\infty \mathcal{C}(r, \tilde{A}(R)) \frac{dR}{\psi(R)} + \left\{ \sum_{\nu=1}^n \int_0^\infty |\mathcal{N}(r, a_\nu(R)) - \mathcal{N}(r, b_\nu(R))| \frac{dR}{\psi(R)} \right\}$$
$$\leq \int \int_{D(r)} \frac{|w'(z)|}{\psi(|w(z)|)} d\sigma \qquad (3.2.7)$$

holds.

The magnitudes in the figure brackets in the left-hand side characterize the Nevanlinna Difference. Therefore Theorem 1 can be considered as an analog of the Second Fundamental Theorem of Nevanlinna, which is valid for all meromorphic functions in the unit disk and in the complex plane. In addition, this theorem, in a sense, measures the closeness between a- and b-points. Thus, Theorem 1 also gives an integrated version of the "Proximity Property of a-points" for all such functions.

3.2.4. Particular cases: functions meromorphic in the complex plane and meromorphic functions with bounded spherical area in the unit disk . Set $\psi(R) = 1 + R^2$, then as a corollary to (3.2.7) we obtain the following.

Theorem 2. *Under conditions of Theorem 1 the following inequality is true:*

$$\int_0^\infty \mathcal{C}(r, \tilde{A}(R)) \frac{dR}{1+R^2} + \sum_{\nu=1}^n \int_0^\infty |\mathcal{N}(r, a_\nu(R)) - \mathcal{N}(r, b_\nu(R))| \frac{dR}{1+R^2}$$
$$\leq \pi r A^{1/2}(r). \tag{3.2.8}$$

First, consider the case of functions meromorphic in the complex plane. Observe that, according to Section 3.1, inequality (3.2.8) gives an integrated closeness for the magnitudes $\mathcal{N}(r, a_\nu(R))$ and $\mathcal{N}(r, b_\nu(R))$. Remember that the closeness of similar magnitudes $N(r, a_\nu)$ and $N(r, b_\nu)$ is the main conclusion of the Nevanlinna Theory. Now, to compare the closeness given by inequality (3.2.8) and by the Nevanlinna Theory let us consider the case when $a_\nu(R)$ and $b_\nu(R)$ lie on some straight lines. Using Nevanlinna's main theorems and Frostman's Theorem, see Nevanlinna [1], one can easily obtain for given ε, $0 < \varepsilon = \text{const} < 1/2$

$$\sum_{\nu=1}^n \int_0^\infty |N(r, a_\nu(R)) - N(r, b_\nu(R))| \frac{dR}{1+|R|^2} = O\left(T^{1/2+\varepsilon}(r)\right), \quad r \to \infty.$$

Similarly, using Ahlfors' First Fundamental Theorem, one can easily obtain

$$\sum_{\nu=1}^n \left| \int_0^\infty (\mathcal{N}(r, a_\nu(R)) - \mathcal{N}(r, b_\nu(R))) \frac{dR}{1+|R|^2} \right| = O\left(rA^{1/2}(r)\right), \quad r \to \infty.$$

Thus we see that all the above conclusions related to the numbers of $a_\nu(R)$ and $b_\nu(R)$-points are quite similar; these numbers must be close to each other.

However, because of the presence $\mathcal{C}(r, \tilde{A}(R))$ in the inequality (3.2.8) we obtain additionally conclusions related to the geometric locations: the $a_\nu(R)$- and $b_\nu(R)$-points must be also geometrically close to each other (Proximity Property), i.e. the average distance $|z_i(a_\nu(R)) - z_i(b_\nu(R))|$ must be a magnitude of the order $rA^{-1/2}(r)$.

Theorem 2'. *Let $w(z)$ belong to the class of functions with bounded spherical area in the unit disk, i.e. let $\lim_{r\to 1} A(r) < C = \text{const} < \infty$.*

Then for the above defined points $a_\nu(R)$ and $b_\nu(R)$, $\nu = 1, 2, \ldots, n$ the following inequality is valid:

$$\int_0^\infty \mathcal{C}(r, \tilde{A}(R)) \frac{dR}{1 + |R|^2}$$

$$+ \sum_{\nu=1}^n \int_0^\infty |\mathcal{N}(r. a_\nu(R)) - \mathcal{N}(r, b_\nu(R))| \frac{dR}{1 + |R|^2} \leq \pi C^{1/2}. \tag{3.2.9}$$

3.2.5. Particular cases: Dirichlet class and functions $w(z)$ with $w'(z)$ from H^1.

According to definitions for functions of Dirichlet class

$$\int\int_{D(1)} |w'(z)|^2 d\sigma < C = \text{const} < \infty,$$

and for the class of functions with $u'(z) \in H^1$

$$\sup \int_0^{2\pi} |w'(z)| d\phi < C = \text{const} < \infty.$$

Set $\psi(R) = 1$, then Theorem 1 implies the following.

Theorem 3. *Let $w(z)$ be from one of the mentioned classes of functions in the unit disk. Then for the above defined points $a_\nu(R)$ and $b_\nu(R)$, $\nu = 1, 2, \ldots, n$,*

$$\int_0^\infty \mathcal{C}(r, \tilde{A}(R))dR + \sum_{\nu=1}^n \int_0^\infty |\mathcal{N}(r. a_\nu(R)) - \mathcal{N}(r. b_\nu(R))| dR \leq C. \tag{3.2.10}$$

3.2.6. Particular cases: bounded functions.

For holomorphic functions bounded in the unit disk from the known Schwarz' Lemma, it immediately follows that

$$\int\int_{D(r)} |w'(z)| d\sigma \leq C \ln \frac{1}{1-r}. \quad C = C(w) = \text{const} < \infty.$$

Set $\psi(R) = 1$, then Theorem 1 implies the following.

Theorem 4. *Let $w(z)$ be a bounded holomorphic function in the unit disk. Then for the above defined points $a_\nu(R)$ and $b_\nu(R)$, $\nu = 1, 2, \ldots, n$,*

$$\int_0^\infty C(r, \tilde{A}(R)) dR$$

$$+ \sum_{\nu=1}^n \int_0^\infty |\mathcal{N}(r, a_\nu(R)) - \mathcal{N}(r, b_\nu(R))| dR \leq C \ln \frac{1}{1-r}. \quad (3.2.11)$$

3.2.7. Particular cases: Blaschke products. For a Blaschke product

$$B(z) := \prod_{(i)} \frac{\bar{z}_i}{|z_i|} \frac{z_i - z}{1 - \bar{z}_i z}$$

with zeros $z_i \in D(1)$ satisfying the conditions $0 < |z_i| < 1$, and

$$\sum_{(i)} (1 - |z_i|) \ln \frac{1}{1 - |z_i|} < \infty \quad (3.2.12)$$

we have

$$\int\int_{D(r)} \left| \frac{B'(z)}{B(z)} \right| d\sigma \leq C = \text{const} < \infty.$$

For proving this it is enough to show that there exists a constant c such that for any $z_i \in D(1)$

$$\int\int_{D(1)} \left| \frac{1}{z - z_i} + \frac{\bar{z}_i}{1 - \bar{z}_i z} \right| d\sigma < c(1 - |z_i|) \ln \frac{1}{1 - |z_i|}.$$

The following beautiful calculation using geometrical reasoning based on the main identity of Section 1.1 was given by Sukiasian. We have

$$\int\int_{D(1)} \left| \frac{1}{z - z_i} + \frac{\bar{z}_i}{1 - \bar{z}_i z} \right| d\sigma$$

$$= \int\int_{D(1)} \left| \frac{f'(z)}{f(z)} \right| d\sigma = \int_0^1 \frac{L(1, \Gamma(R), f)}{R} dR,$$

where $f(z) = (z - z_i)/(1 - \bar{z}_i z)$. Note that

$$\left\{ z : \left| \frac{z - z_i}{1 - \bar{z}_i z} \right| = R \right\}$$

$$= \left\{ z : \left| z - \frac{z_i(1 - R^2)}{1 - |z_i|^2 R^2} \right| = \frac{R(1 - |z_i|^2}{1 - |z_i|^2 R^2} \right\}$$

so that

$$L(1, \Gamma(R), f) = \frac{2\pi R(1 - |z_i|^2}{1 - |z_i|^2 R^2},$$

and consequently

$$\iint_{D(1)} \left| \frac{1}{z - z_i} + \frac{\bar{z}_i}{1 - \bar{z}_i z} \right| d\sigma = \int_0^1 \frac{L(1, \Gamma(R), f)}{R} dR$$

$$= \pi \frac{1 - |z_i|^2}{|z_i|} \ln \frac{1 + |z_i|}{1 - |z_i|}.$$

Now if we take $\psi(R) = R$, then Theorem 1 implies the following.

Theorem 5. *Let $w(z)$ be a Blaschke product with zeros satisfying the condition (3.2.12). Then for the above defined points $a_\nu(R)$ and $b_\nu(R)$, $\nu = 1, 2, \ldots, n$, there exists a constant $C < \infty$ such that*

$$\int_0^\infty C(r, \tilde{A}(R)) \frac{dR}{R} + \sum_{\nu=1}^n \int_0^\infty |\mathcal{N}(r, a_\nu(R)) - \mathcal{N}(r, b_\nu(R))| \frac{dR}{R} \leq C. (3.2.13)$$

Let us once more return to the above idea of sufficiently large sets. If we take a big number n of pairs $a_\nu(R)$ and $b_\nu(R)$, then from (3.2.13) we conclude that for the majority of ν and R the magnitudes

$$|\mathcal{N}(r, a_\nu(R)) - \mathcal{N}(r, b_\nu(R))|$$

are "small", and consequently the majority of pairs of $a_\nu(R)$- and $b_\nu(R)$-points occur in the sets $\tilde{A}(R)$ and again by (3.2.13) they are close to one another. Since

$$\int_0^\infty C(r, \tilde{A}(R)) \frac{dR}{R}$$

$$= \int_0^\infty \sum_{(z_j(a_\nu(R)), z_j(b_\nu(R))) \in \tilde{A}(R)} |z_j(a_\nu(R)) - z_j(b_\nu(R))| \frac{dR}{R} \leq C,$$

we can qualitatively conclude that on average

$$\sum_{(z_j(a_\nu(R)),z_j(b_\nu(R)))\in\tilde{A}(R)} |z_j(a_\nu(R)) - z_j(b_\nu(R))| \le C_1 = \text{const} < \infty.$$

Observe that the condition (3.2.12) is very weak and consequently the estimate (3.2.13) holds for large classes of Blaschke products.

It would be interesting to find the estimates of the type

$$\int\int_{D(r)} |B'(z)|d\sigma \le b(r), \tag{3.2.14}$$

where $B(z)$ is an arbitrary Blaschke product, $b(r)$ is given in terms of the zeros of $B(z)$. Then one can obviously use Theorem 6 with $\psi(R) = 1$ and obtain some more simple conclusions similar to Theorem 10 for arbitrary Blaschke products.

3.2.8. Particular cases: functions $w(z)$ from Hardy classes H^p. By the Hardy–Stein–Spencer identity (see Section 1.1) for any function $w(z)$ holomorphic in the unit disk

$$rJ_p'(r) = \frac{p^2}{2\pi}\int\int_{D(r)} \frac{|w'(z)|^2}{|w(z)|^{2-p}}d\sigma,$$

where

$$J_p(r) = \frac{1}{2\pi}\int_0^{2\pi} |w(re^{i\theta}|^p d\theta.$$

Hence applying the Cauchy–Buniakovski inequality we derive

$$\int_0^\infty L(r,\Gamma(R))\frac{dR}{R^{(2-p)/2}} = \int\int_{D(r)} \frac{|w'(z)|}{|w(z)|^{(2-p)/2}}d\sigma \le \frac{\pi}{p}[2r^3 J_p'(r)]^{1/2}.$$

Consequently

$$\int_0^r \int_0^\infty \frac{L(r,\Gamma(R))}{R^{(2-p)/2}}dRdr \le C(p)[J_p(r)]^{1/2},$$

where $C(p)$ is a constant depending only on p. Thus, for any function $w \in H^p$, $0 < p < \infty$, there exists a constant $C(p,w)$ depending only on p and w such that

$$\int_0^r \int_0^\infty \frac{L(r,\Gamma(R))}{R^{(2-p)/2}}dRdr \le C(p,w).$$

Now applying Theorem 1 with $\psi(R) = R^{(2-p)/2}$ we come to the following.

Theorem 6. *Let $w(z) \in H^p$, $0 < p < \infty$. Then for the above defined points $a_\nu(R)$ and $b_\nu(R)$, $\nu = 1, 2, \ldots, n$, the following inequality*

$$\int_0^r \int_0^\infty C(r, \tilde{A}(R)) \frac{dRdr}{R^{(2-p)/2}}$$
$$+ \sum_{\nu=1}^n \int_0^\infty \int_0^r |\mathcal{N}(r, a_\nu(R)) - \mathcal{N}(r, b_\nu(R))| \frac{dRdr}{R^{(2-p)/2}} \le C(p, w)$$

(3.2.15)

holds.

3.2.9. Remarks on the role of double integrals in this section.

Nevanlinna's counting function $N(r, a)$ of a-points is usually estimated by the characteristic function $T(r, w)$ or by the spherical characteristic function $A(r, w)$ if we study functions meromorphic in the complex plane (or in the unit disk, under the additional condition of fast growth). In this section, we describe the closeness of a-points for the same classes of functions in terms of

$$A_1(r, w) = \int \int_{D(r)} \frac{|w'(z)|}{1 + |w(z)|^2} d\sigma.$$

From the proof of the Proximity Property [12, 13] we conclude that the smaller is $A_1(r, w)$ (in comparison with $A(r, w)$) and the stronger is the closeness of a-points.

The smallness of $A_1(r, w)$ plays a crucial role in the "marvelous" Littlewood's Property of entire functions of finite order stating that the majority of a-points for the majority of values a occurs in some subsets of the complex plane with comparatively small area. Littlewood also conjectured that $A_1(r, w)$ is strongly less than $A^{1/2}(r, w)$ for polynomials and he has derived his Property under the assumption that this conjecture is true. Littlewood's Property and conjecture are in a good accord with the above conclusions that the smallness of $A(r, w)$ leads to the closeness of a-points. However, the last assertion embraces incomparably large classes of functions.

By the way, if Littlewood's conjecture is true, then the Proximity Property gives a more detailed description of Littlewood's Property, (this will be published somewhere else).

Assertions obtained in this section evidently show that for meromorphic functions with "slow" growth the counting functions $\mathcal{N}(r, a)$, as well as the closeness of a-points, are described by the function $A_1(r, w)$ (but not $A(r, w)$) or by the above double integrals with different weights corresponding to various functions $\psi(R)$. Consequently, in the case of functions with slow growth the above double integrals play the role of characteristic functions in description of the quantities and the closeness of a-points.

CHAPTER 4

DISTRIBUTION OF Γ-LINES FOR FUNCTIONS MEROMOR-PHIC IN ℂ: APPLICATIONS

In the Introduction and at the end of Chapter 1 we studied Γ-lines of various classes of functions and gave general approaches to the investigation of various problems. This chapter is devoted to the development of the mentioned approaches for functions meromorphic in ℂ. The main results and their immediate applications were attained in the author's paper [9]. Other applications using ideas of Γ-lines one can find in author's papers [7],[8],[10],[11].

4.1. The Main Results on Distribution of Γ-lines

In these subsections we state some theorems which will be proved in the later ones.

4.1.1. The Tangent Variation Principle (see Section 1.1) gives a detailed information about Γ-lines for wide classes of functions. For a definite class of function it is appropriate to find the estimates of lengths of Γ-lines and describe the distribution of these lines in terms of characteristics of the same classes. This can be successfully done for functions meromorphic in ℂ and leads to the following results.

Theorem 1. *Let* $w(z) \in M$ *and let* Γ *be any smooth Jordan curve (bounded or unbounded) for which* $\nu(\Gamma) < \infty$.[9] *Then there exist absolute constants* $K < \infty$ *and* $C \in (0, 1)$ *such that*

$$L(r, \Gamma) \leq KrA(r), \quad r \in E. \tag{4.1.1}$$

where $L(r, \Gamma) := L(D, \Gamma)$, $D = \{z : |z| < r\}$, *and* E *is some set of lower logarithmic density* C.

If $c = \text{const} > 1$, *then for* $r > r_0$

$$L(r, \Gamma) \leq K(c, \Gamma) r T(c, r), \tag{4.1.1'}$$

---

[9]The full variation of the angle between the tangent to Γ and the real axis is finite (see Section 1.2.3).

where $K(c, \Gamma) = $ const $< \infty$.

The next theorem is our main result on the distribution of Γ-lines.

Theorem 2. *Let $w(z) \in M$ and let Γ_ν, $\nu = 1, \ldots, q$, be a collection of disjoint, bounded, smooth Jordan curves with $\nu(\Gamma_\nu) < \infty$.*

Then there exist absolute constants $K < \infty$ and $C \in (0,1)$ such that

$$\sum_{\nu=1}^{q} L(r, \Gamma_\nu) \leq KrA(r), \quad r \in E, \tag{4.1.2}$$

where E is some set of lower logarithmic density C.

This theorem shows that there is a definite analogy between the behaviors of the magnitude $L(r, \Gamma_\nu)$ and Nevanlinna's functions $m(r, a_\nu)$.

One can rewrite the inequality (4.1.2) in the form

$$\sum_{\nu=1}^{q} \Delta(\Gamma_\nu) \leq K \tag{4.1.3}$$

where

$$\Delta(\Gamma) := \liminf_{r \to \infty} \frac{L(r, \Gamma)}{rA(r)}.$$

If q is "large", then for the majority of ν the magnitudes $\Delta(\Gamma_\nu)$ are "small" ($\leq K/q$ in average). Letting $q \to \infty$ one can observe that the typical (i.e. not "deficient") curves Γ_ν are those curves Γ_ν for which $\Delta(\Gamma_\nu) = 0$. Accordingly, we shall call $\Delta(\Gamma)$ "deficiency" of Γ and say that Γ is a "deficient" or "not deficient" curve if $\Delta(\Gamma) > 0$ or $\Delta(\Gamma) = 0$ respectively.

Theorem 2 implies a corollary on the distribution of Γ-lines, similar to the Nevanlinna deficiency relation. For stating this corollary we denote by $M(\Gamma)$ a set $\{\Gamma\}$ of bounded, disjoint smooth Jordan curves and adopt $M(\Gamma(R))$ for the particular case when $\{\Gamma\} = \{\Gamma(R)\}$ is the set of concentric circles with the radii R.

Corollary 1 (Deficiency Relation for Curves). *Let $w(z) \in M$. Then for any curve $\Gamma \in M(\Gamma)$, with the possible exception of not more than countably many curves every $\Gamma_\mu \in M(\Gamma)$ satisfies*

$$\Delta(\Gamma) = 0, \tag{4.1.4}$$

i.e. the deficiencies of these Γ are equal to zero. Besides, for deficient curves Γ$_ν$

$$\sum_{\mu} \Delta(\Gamma_\mu) \leq K, \tag{4.1.5}$$

where $K < \infty$ is an absolute constant.

4.1.2. The magnitudes of the deficiencies $\Delta(\Gamma)$ of curves Γ show the growth of the length $L(r, \Gamma)$ on a sequence $r = r_n \to \infty$. To estimate L for all r it would be natural to consider the magnitude

$$\widetilde{\Delta}(\Gamma) = \lim_{r \to \infty} \sup \frac{L(r, \Gamma)}{rA(r)}. \tag{4.1.6}$$

In the theory of distribution of Γ-lines this magnitude plays the same role as Valiron's deficient value

$$\widetilde{\delta}(a) = \lim_{r \to \infty} \sup \frac{m(r, a)}{T(r)}$$

plays in Value Distribution Theory. Recall (see Section 2.1.4) that $\widetilde{\delta}(a) = 0$ for the majority of values of $a \in \overline{\mathbb{C}}$ (exactly. for all $a \in \overline{\mathbb{C}}$ with possible exception of a set of zero capacity).

The following result is an analog of the last statement for the theory of Γ-lines.

Theorem 3. *Let $w(z) \in M$ and let $\varphi(r)$ be a positive, monotonic function on $(0, \infty)$, such that $\varphi(r) \to 0$ as $r \to \infty$. Then for any $R > 0$, with the possible exception of a set of measure zero,*

$$\lim_{\rho \to \infty} \sup \frac{L(r. \Gamma(R))\varphi(r)}{rA^{1/2}(r)} = 0. \tag{4.1.7}$$

Thus, choosing $\varphi(r) > A^{-\varepsilon}(r)$. $\varepsilon < 1/2$. we obtain

$$\widetilde{\Delta}(\Gamma(R)) = 0$$

for those R for which (4.1.7) is valid.

4.1.3. It is known that if the Riemann surface of the function w^{-1} over a domain D has a "proper" structure then the magnitude $m(r, a)$, $a \in D$, is "small" (see Kakutany [1], Selberg [1], [2], Tumura [1]). The following is a similar conclusion for the theory of Γ-lines.

Theorem 4. *Let $w(z) \in M$, let D be a given domain which lies with its boundary in a domain $D' \subset \mathbb{C}$ and let Γ be a smooth Jordan curve with $\nu(\Gamma) < \infty$ which lies completely in D.*

If the Riemann surface of the function w^{-1} is not ramified over D', then there exists a constant K depending only on the structures of Γ, D, and D' such that

$$L(r, \Gamma) \leq K r A^{1/2}(r), \quad r_0 < r < \infty. \tag{4.1.8}$$

4.1.4. Sharpness of the estimates of L. First we note that if $w(z) = e^z$ and Γ is the segment $[0, 1]$, then there exists a constant $C > 0$ such that

$$L(r, \Gamma) > CrA(r).$$

Thus, the estimates of Theorems 1 and 2 are precise up to constants for some functions. The following theorem indicates a class of such functions. We use the concept of geometric deficient values $\overline{\delta}(a)$ introduced in Section 2.2.

Theorem 5. *Let $w(z)$ be any function of M_R, let $\overline{\delta}(a) = \overline{\delta} > 0$ for $a \in \mathbb{C}$, and let Γ be a continuous curve in the w-plane, such that a belongs to the closure of Γ. Then for any fixed $\varepsilon > 0$*

$$L(r, \Gamma) > (\overline{\delta} - \varepsilon) \int_0^r A(t)dt - h \int_0^r L(t)dt, \quad r < R, \tag{4.1.9}$$

where $h \in [0, \infty)$ is a constant. Besides, if Γ is a straight line, then

$$L(r, \Gamma) > (\overline{\delta} - \varepsilon) \int_0^r A(t)dt. \tag{4.1.10}$$

Using Lemma 5 of Section 2.3 we get

$$
\begin{aligned}
L(r, \Gamma) &> (\overline{\delta} - \varepsilon) \int_{r/2}^r A(t)dt + o(rA(r)) \\
&\geq 2^{-1}(\overline{\delta} - \varepsilon) A\left(\frac{r}{2}\right) r + o(rA(r)), \quad r > r_0.
\end{aligned}
$$

Thus, by Lemma 6 of Section 2.4.3 we obtain that for $\lambda_w < \infty$ there exists a constant $K(\lambda_w) > 0$ depending only on λ_w and a sequence $r_n \to \infty$ such that

$$L(r, \Gamma) > K(\lambda_w)(\bar{\delta} - \varepsilon) r A(r), \quad r_n \to \infty,$$

i.e. as $r \to \infty$ on some sequences, the above given upper and lower bounds of $L(r, \Gamma)$ differ only by constant multipliers.

One can easily check the sharpness of the estimate (4.1.7) by the Weierschtrass doubly-periodical function $\mathcal{P}(z)$. Indeed, let for a given R a fragment of the line $\Gamma(R)$ contained in a period parallelogram of the function $\mathcal{P}(z)$ be of the length $K(R)$. Since in other parallelograms the geometrical picture is the same, we have $L(r, \Gamma(R)) \sim P(r)K(R)$ where $P(r)$ is the number of parallelograms having common points with $D(r)$. As $P(r) \sim \text{const } r^2$ and $rA^{1/2}(r, P) = O(r^2)$ we conclude that the estimate (4.1.8) is precise up to the constant K and the estimate (4.1.7) is precise up to an arbitrarily slow decreasing function $\varphi(r)$.

4.1.5. Proof of Theorems 1 and 2. Using the inequality (1.3.1') of the Second Fundamental Theorem (see 1.3.2) in the case $D = D(r) = \{z : |z| < r\}$, taking into account (4.1.9) of Section 1.2.2 and that $A_1(D) = \int_0^r L(t)dt$ and $l(D) = 2\pi r$ we obtain the following estimate

$$L(r, \Gamma) \leq 2K \int\int_{|z| < r} \left| \frac{w''(z)}{w'(z)} \right| d\sigma + h(\Gamma) \int_0^r L(t)dt + 3\sqrt{8}\pi r. \quad (4.1.11)$$

Now applying the inequalities of Lemmas 1' and 5 of Section 2.4 to the right-hand side of (4.1.11) we come to the estimate (4.1.1) of Theorem 1.

Applying the inequality (1.2.16) of the First Fundamental Theorem (Section 1.2.3) we similarly find

$$L(r, \Gamma) \leq 2K(\Gamma) \int\int_{|z| < r} \left| \frac{w''(z)}{w'(z)} \right| d\sigma + K(\Gamma)2\pi r.$$

Hence using Lemma 1 of Section 2.3 we derive the estimate (4.1.1') of Theorem 1.

Now observe that the inequality (4.1.1) of the Second Fundamental Theorem (Section 1.3.2) implies

$$\sum_{\nu=1}^{q} L(r, \Gamma_\nu) \leq 2K \int\int_{|z| < r} \left| \frac{w''(z)}{w'(z)} \right| d\sigma$$

$$+h(\Gamma_1, \Gamma_2, \ldots, \Gamma_q) \int_0^r L(t)dt + \sqrt{8}\pi r. \quad (4.1.11')$$

Hence, we obtain Theorem 2 arguing similarly as in the above proof of the inequality (4.1.1) of Theorem 1.

4.1.6. Proof of Theorem 3. Using the modification of the length–area principle given by means of the inequality (1.1.3) of Section 1.1.2 for the domain $D = D(r)$ we get

$$\int_0^\infty \frac{L(r, \Gamma(R))}{1 + R^2} dR \le \pi r A^{1/2}(r).$$

As the expression

$$\frac{L(r, \Gamma(R))\varphi(r)}{(1 + R^2)r A^{1/2}(r)}$$

represents a function continuous in R, its upper limit as $r \to \infty$ is a Borel measurable function. Therefore we are allowed to let $r \to \infty$ under the integral and hence obtain

$$\int_0^\infty \lim_{r\to\infty} \sup \frac{L(r, \Gamma(R))\varphi(r)}{(1 + R^2)r A^{1/2}(r)} dR \le \lim_{r\to\infty} \sup \pi\varphi(r) = 0,$$

i.e. the assertion of Theorem 3 follows.

4.1.7. Proof of Theorem 4. It immediately follows from Lemmas 1 and 2 of Section 2.3 if we take into account that under the conditions of the theorem inequality (2.3.8) of Section 2.3 holds.

4.1.8. Proof of Theorem 5. It is based on the below inequality.

Lemma 1. *Under the conditions of Theorem 5*

$$\Phi(r, \Gamma) \ge (\bar{\delta} - \varepsilon)A(r) - h(a, \Gamma)L(r), \quad r > r_0, \quad (4.1.12)$$

where $h(a, \Gamma) = \mathrm{const} \in [0, \infty)$ *and* $\Phi(r, \Gamma)$ *is the number of the points* $z_i \in \{z : |z| = r\}$ *at which* $w(z_i) \in \Gamma$. *Particularly, if* Γ *is a straight line, then*

$$\Phi(r, \Gamma) \ge (\bar{\delta} - \varepsilon)A(r). \quad (4.1.13)$$

Proof. First we suppose $a \neq \infty$ and consider two cases.

(a) Let $\Gamma \bigcap \{w : |w - a| \geq 1\} \neq \emptyset$. It is easy to see that $\Phi(r, \Gamma) \geq \nu(r, a)$ if the whole boundary ∂F_r lies over the disk $|w - a| \leq 1$. This follows from the geometric interpretation of the magnitude $\nu(r, a)$ as the total number of complete turns of ∂F_r around the point a and from the fact that ∂F_r itself is an arc γ_a, i.e. $\nu(r, a) = [\nu_{\gamma_a}]'$ in the considered case. If ∂F_r has common points with $|w - a| > 1$, then the length of any arc γ_a for which $|[\nu_{\gamma_a}]'| \geq 1$ is not less than 1 and such an arc is intersected with the curve Γ in at least $|[\nu_{\gamma_a}]'| - 2$ points. Denoting the length of an arc γ_a by l_{γ_a} we get

$$\Phi(r, \Gamma) \geq \nu(r, a) - 2 \sum_{\gamma_a} l_{\gamma_a}.$$

Further, denoting the spheric length of the image of γ_a by L_{γ_a} and observing that $l_{\gamma_a} < h(a) L_{\gamma_a}$, where $h(a) = \text{const} \in [0, \infty)$, we obtain

$$\Phi(r, \Gamma) \geq \nu(r, a) - 2h(a) \sum_{\gamma_a} L_{\gamma_a} \geq \nu(r, a) - hL(z). \qquad (4.1.14)$$

Taking into account that $\overline{\delta}(a) = \overline{\delta} > 0$ we come to (4.1.12).

(b) Let now $\Gamma \subset \{w : |w - a| < R_0\}$. $R_0 < 1$. Suppose that $|[\nu_{\gamma_a}]'| \geq 1$ for an arc γ_a and the set $\{z : |z| = r, \; \varphi_1 \leq \varphi \leq \varphi^*\}$ is the pre-images of this arc. We separate in γ_a the arcs γ^i successively as φ increases from φ_1 to φ^* such that the increase of $\arg(w(z) - a)^{-1}$ on γ^i is equal to 2π or -2π. The pre-images of the set of arcs γ^i is some set $\{z : |z| = r, \; \varphi_1 \leq \varphi \leq \varphi'\}$, $\varphi' \leq \varphi^*$. Let γ' be the set of those γ^i for which $\gamma^i \bigcap \{w : |w - a| = R_0\} \neq \emptyset$ and γ^0 be the w-image of the set $\{z : |z| = r, \; \varphi' \leq \varphi \leq \varphi^*\}$. Then the set $\gamma'' := \gamma_a \setminus \{\gamma' \cup \gamma^0\}$ is the sum of some set of arcs $\gamma_j'' \subset \gamma''$. Denote the sum of γ_j'' and its neighboring arcs $\gamma^i \subset \gamma'$. or possibly γ^0, by $\overline{\gamma}_j''$; denote the number of arcs γ^i from γ' or γ_j'' by τ' or τ_j'' respectively; and denote the length of arcs $\gamma^i \subset \gamma'$ and $\gamma_i \subset \overline{\gamma}_j''$ by $l_{\tau_i'}$ and $l_{\overline{\gamma}_j''}$ respectively. As $l_{\tau_i'} > R_0$ we deduce

$$\tau' < \frac{1}{R_0} \sum_{\tau'} l_{\tau_i'}.$$

It is evident that the number of intersections of an arc γ_j'' with the curve Γ at least is $\tau_j'' - 2$, whence taking into account that $l_{\overline{\gamma}_j''} > R_0$ we get

$$\Phi(r, \Gamma) \geq \sum_{\gamma_a} \sum_j \tau_j'' - \frac{2}{R_0} l_{\overline{\gamma}_j''}.$$

The last two inequalities imply

$$\Phi(r,\Gamma) \geq \sum_{\gamma_a} \left(\sum_J \tau_j'' + \tau' \right) - \sum_{\gamma_a} \left(\frac{2}{R_0} \sum_j l_{\overline{\gamma}_j''} + \frac{1}{R_0} \sum_{\tau'} l_{\tau_i'} \right)$$
$$\geq \nu(r,a) - hL(r),$$

whence (4.1.12) follows as a $\overline{\delta}(a) = \overline{\delta} > 0$.

For deriving the inequality (4.1.12) in the case $a = \infty$ one can repeat the above arguments for the function $W(w)$ and for the curve $W(\Gamma)$, where W is a linear-fractional transformation which rotates the sphere and does not change the quantities Φ, ν and L.

The inequality (4.1.13) immediately follows from the definition of $\overline{\delta}(a)$ and from the conditions of Theorem 5. Thus, our lemma is completely proved.

For proving Theorem 5 we suppose that l is a bounded, smooth curve in the z-plane, $\Psi(r,l)$ is the number of its intersections with the circle $|z| = r$ and $|l|$ is the length of l. Further, we put a point r' from the real axis in correspondence with any point $z \in l$ ($z = r'e^{i\varphi}$), so that we allow the possibility for different points $z \in l$ to have the same projection r'. Assuming that $\Psi(r,l)$ is finite for any r we conclude that the curve l can be divided into parts l_i, in a way providing a one-to-one correspondence between l_i and the intervals p_i from the set of all r' corresponding to l_i. Let $|p_i|$ be the length of p_i, then obviously

$$|l| > \sum_i |p_i|.$$

Evidently, this inequality remains valid even if it is not assumed that $\Psi(r,l)$ is finite. Further, by the geometric meaning of an integral

$$\sum_i |p_i| = \int_{\sigma_2(l)}^{\sigma_1(l)} \Psi(r,l)dr,$$

where $\sigma_1(l) = \max_{z \in l} |z|$, $\sigma_2(l) = \min_{z \in l} |z|$. From the last two relations for the set of curves l_i we find

$$\sum_j |l_j| \geq \sum_j \int_{\sigma_2(l_j)}^{\sigma_1(l_j)} \Psi(r,l)dr. \tag{4.1.15}$$

Now let the set of curves l_j be $w^{-1}(\Gamma) \bigcap \{z : |z| < r\}$, where Γ and $w(z)$ are those defined in Theorem 5 of Section 4.1.4, and let as above $\Phi(r,\Gamma)$ be the number of those points z_i on $|z| = r$, for which $w(z_i) \in \Gamma$. As

$$\Phi(r, \Gamma) = \sum_j \Psi(r, l_j) \quad \text{and} \quad L(r, \Gamma) = \sum_j |l_j|,$$

using (4.1.12), (4.1.13) and (4.1.15) of Lemma 1 we derive the inequalities (4.1.9) and (4.1.10) of Theorem 5.

4.2. "Windings" of Γ-lines

4.2.1. First we shall introduce a magnitude $\mathcal{L}(r, \Gamma)$ characterizing the disposition of Γ-lines in the disk $|z| < r$ with respect to their winding around the origin.

Let $\{\gamma_i(t)\}$, $t \in [0, 1]$, be a set of smooth curves, the sum of which coincides with the set $w^{-1}(\Gamma) \cap D(r)$, where $D(r) = \{z : |z| < r\}$. Further, let $\mathcal{L}(\gamma_i)$ be the complete variation of the argument of a curve γ_i, i.e.

$$\mathcal{L}(\gamma_i) := \int_0^1 \left| \frac{\partial}{\partial t} \arg \gamma_i(t) \right| dt,$$

and

$$\mathcal{L}(r, \Gamma) := \sum_i \int_0^1 \left| \frac{\partial}{\partial t} \arg \gamma_i(t) \right| dt := \sum_i \mathcal{L}(\gamma_i).$$

Using clear connections between the length of an arc in the ring $\{r/2 < |z| < r\}$ and the angle in which this arc is seen from the origin, one can observe that

$$\mathcal{L}(r, \Gamma) \leq \sum_{n=0}^{[\log_2 r]'} \left\{ \mathcal{L}\left(\frac{r}{2^n}, \Gamma\right) - \mathcal{L}\left(\frac{r}{2^{n+1}}, \Gamma\right) \right\} + O(1) < K A(r) \log_2 r + O(1)$$

as $r \in E$, where $[x]'$ is the entire part of x. The following theorem which we shall prove in Section 4.2.2 shows that the above estimate of the winding of $\mathcal{L}(r, \Gamma)$ can be substantially sharpened.

Theorem 1. *Let $w(z) \in M$ and let Γ_ν, $\nu = 1, 2, \ldots, q$, be some disjoint Jordan curves with $\nu(\Gamma_\nu) < \infty$, such that $\Gamma_1, \ldots, \Gamma_{q-1}$ are bounded, but Γ_q is unbounded. Then*

$$\sum_{\nu=1}^q \mathcal{L}_i(r, \Gamma_\nu) \leq K(c)\widetilde{T}(cr, w) \quad as \quad r > r_0, \qquad (4.2.1)$$

where $K(c)$ is a constant depending solely on $c = \text{const} > 1$ and

$$\widetilde{T}(r) = \int_0^r \frac{T(t)}{t} dt.$$

Obviously from (4.2.1) a similarity of the deficiency relation follows. We have omitted it. On the other hand, Theorem 1 implies some obvious conclusions, particularly indicating that asymptotic Γ-lines cannot revolve "very strongly" around the origin, in general.

4.2.2. Proof of Theorem 1. First we shall establish the general inequality

$$\sum_{\nu=1}^q \mathcal{L}(r, \Gamma_\nu) \leq \frac{32}{\pi} \int_0^r \int_0^{2\pi} \left| \frac{\partial}{\partial t} \arg \frac{\partial}{\partial t} w(z) \right| dt d\varphi$$

$$+ \frac{64\sqrt{2}}{\pi} \int_0^r \int_0^{2\pi} \left| \frac{\partial}{\partial \varphi} \arg \frac{\partial}{\partial \varphi} w(z) \right| d\varphi \frac{dt}{t} +$$

$$+ h\left(\Gamma_1, \Gamma_2, \ldots, \Gamma_q \right) \int_0^r \int_0^{2\pi} \frac{|w'(z)|}{1 + |w(z)|^2} dt d\varphi + 4 \ln r + 8\pi + h$$

$$= T_1 + T_2 + T_3 + 4 \ln r + 8\pi + h, \tag{4.2.2}$$

where $h = h\left(\Gamma_1, \Gamma_2, \ldots, \Gamma_q \right) = \text{const} < \infty$. For proving (4.2.2) suppose that l_φ is a set of some smooth curves $\Psi_i(t)$, $t \in [0,1]$, such that (a) $\Psi_i(t) \in w^{-1} \left(\Gamma_1 \cap \Gamma_2 \cap \ldots \cap \Gamma_q \right) \cap \{|z| \leq r\}$, (b) for any point $z_0 \in \Psi_i(t)$ the smaller angle between the straight line $\{z : \arg z = \arg z_0\}$ and $\Psi_i(t)$ is at least $\pi/4$. Further, we suppose that l_r is a set of some smooth curves $\Psi_i^*(t)$, $t \in [0,1]$, such that (a) $\Psi_i^*(t) \in w^{-1} \left(\Gamma_1 \cap \Gamma_2 \cap \ldots \cap G_q \right) \cap \{|z| \leq r\}$, (b) for any point $z_0 \in \Psi_i^*(t)$ the smaller angle between the circle $\{z : |z| = |z_0|\}$ and $\Psi_i^*(t)$ is greater than $\pi/4$. Also we denote

$$\mathcal{L}(l_\varphi) = \sum_i \int_0^1 \left| \frac{\partial}{\partial t} \arg \Psi_i(t) \right| dt, \quad \mathcal{L}(l_r) = \sum_j \int_0^1 \left| \frac{\partial}{\partial t} \arg \Psi_j^*(t) \right| dt.$$

Obviously

$$\sum_{\nu=1}^q \mathcal{L}(r, \Gamma_\nu) = \mathcal{L}(l_\varphi) + \mathcal{L}(l_r). \tag{4.2.3}$$

Further, we denote $J_\varphi = \{z : |z| \leq r, \ \arg z = \varphi\}$, $D(r) = \{z : |z| < r\}$, and suppose that Ψ_φ is the set of common points of J_φ and l_φ, and $n(\Psi_\varphi)$

is the number of these points. Similarly we suppose that Ψ_r^* is the set of common points of $\partial D(r)$ and l_r, and $n(\Psi_r^*)$ is the number of these points. Then obviously

$$L(l_\varphi) = \int_{[0,2\pi]\backslash\{\varphi^*\}} n(\Psi_\varphi)d\varphi = \int_0^{2\pi} n(\Psi_\varphi)d\varphi,$$

where $\{\varphi^*\}$ is the set of those φ for which there is at least one singular point of $w(z)$ on J_φ. For any not singular point $z_0 \in \Psi_i(t)$ the smaller angle between the w-image $J_{\arg z_0}$ and Γ_ν containing $w(z_0)$ by conformality is not less than $\pi/4$ at the point w_0. Hence, by the above definitions

$$n(\Psi_\varphi) = \sum_{\nu=1}^{q} \Phi_{\pi/4}(J_\varphi, \Gamma_\nu)$$

for $\varphi \in [0, 2\pi]\backslash\{\varphi^*\}$ (for definition of $\Phi_a(\mu, \Gamma)$ see Section 1.3.2), whence

$$\mathcal{L}(l_\varphi) = \int_0^{2\pi} \sum_{\nu=1}^{q} \Phi_{\pi/4}(J_\varphi, \Gamma_\nu)d\varphi. \qquad (4.2.4)$$

Consequently, by Lemma 1 of Section 1.3.2

$$L(l_\varphi) \leq \frac{32}{\pi} \int_0^r \int_0^{2\pi} \left| \frac{\partial}{\partial t} \arg \frac{\partial}{\partial t} w(z) \right| d\varphi dt$$
$$+ h\left(\Gamma_1, \Gamma_2, \ldots, \Gamma_q\right) \int_0^{2\pi} \int_0^r \frac{|w'(z)|}{1 + |w(z)|^2} dt d\varphi + 8\pi. \qquad (4.2.5)$$

Note that here we used Lemma 1 of Section 1.3.2 in the following way: inequality (1.3.2) was used for the set of curves $\Gamma_1, \Gamma_2, \ldots, \Gamma_{q-1}$ (their contribution in the above coefficient $32/\pi$ is $8/\pi$), and (1.3.2′) was used for Γ_q (the contribution is $24/\pi$).

Now we turn to the evaluation of $\mathcal{L}(l_r)$. Let $\Psi(t)$, $t \in [0, 1]$, be a smooth curve in the ring $r/2 \leq |z| \leq r$, then obviously

$$\int_0^1 \left| \frac{\partial}{\partial t} \arg \Psi(t) \right| dt \leq \frac{2}{r}|\Psi|,$$

where we set $|X|$ as the length of a curve X. Summing up these inequalities over all curves from $l_r \bigcap \{z : r/2 \leq |z| < r\}$ we find

$$\mathcal{L}(l_r \cap D(r)) - \mathcal{L}(l_r \cap D(r/2)) \leq \frac{2}{r}\left(|l_r \cap D(r)| - |l_r \cap D(r/2)|\right).$$

By the standard arguments on connection of the length of a curve and the length of its projection we get

$$\frac{2}{r}\left(|l_r \cap D(r)| - |l_r \cap D(r/2)|\right)$$

$$\leq \frac{2\sqrt{2}}{r}\int_{r/2}^r n(\Psi_t^*)dt \leq 2\sqrt{2}\int_{r/2}^r \frac{n(\Psi_t^*)}{t}dt.$$

Let now $n_0(r)$ be a number such that $1 \leq r2^{-n_0(r)} < 2$. Repeating our argument for the segment $[r2^{-n}, r2^{-(n+1)})$ and by the use of the last inequality we derive

$$\mathcal{L}(l_r \cap D(r)) - \mathcal{L}(l_r \cap D(r2^{-n_0(r)})) \leq 2\sqrt{2}\int_{r2^{-n_0(r)}}^r \frac{n(\Psi_t^*)}{t}dt,$$

or which is the same as

$$\mathcal{L}(l_r) \leq 2\sqrt{2}\int_1^r \frac{n(\Psi_t^*)}{t}dt + h,$$

where $h = \text{const} < \infty$. For any not singular point $z_0 \in \Psi_j^*(t)$ the smaller angle between the w-image of the circle $\partial D(r)$ and Γ_ν containing the point $w(z_0)$ is not less than $\pi/4$ at $w(z_0)$. Hence

$$n(\Psi_t^*) = \sum_{\nu=1}^q \Phi_{\pi/4}(\partial D(t), \Gamma_\nu),$$

in the case when $D(t)$ does not contain singular points. Consequently,

$$\mathcal{L}(l_r) \leq 2\sqrt{2}\int_1^r \sum_{\nu=1}^q \Phi_{\pi/4}(\partial D(t), \Gamma_\nu)\frac{dt}{t} + h. \tag{4.2.6}$$

Applying here Lemma 1 of Section 1.3.2 similar to the proof of (4.2.5) we get

$$\mathcal{L}(l_r) \leq \frac{64\sqrt{2}}{\pi}\int_1^r \int_0^{2\pi} \left|\frac{\partial}{\partial\varphi}\arg\frac{\partial}{\partial\varphi}w(z)\right| d\varphi\frac{dt}{t}$$

$$+ h(\Gamma_1, \Gamma_2, \ldots, \Gamma_q)\int_1^r \int_0^{2\pi} \frac{|w'(z)|}{1 + |w(z)|^2}dtd\varphi + 4\ln r + h.$$

$$\tag{4.2.7}$$

Now the estimate (4.2.2) follows from (4.2.3), (4.2.5) and (4.2.7). For the evaluation of T_1, T_2 and T_3 on the right-hand side of (4.2.2) we make use of Lemmas 2, 3, and 4 respectively. Inserting these estimates of T_1, T_2 and T_3 into (4.2.2) we come to the desired inequality (4.2.1) of Theorem 1.

4.3. Average Lengths of Γ-lines Along Concentric Circles and the Deficient Values

4.3.1. Consideration of the integral mean

$$\int_0^1 \frac{L(r, \Gamma(R, a))}{R} dR,$$

where $\Gamma(R, a) = \{w : |w - a| = R\}$ was offered by Sukiasian. Perhaps, one of the most interesting questions here is: *whether the deficiency of the value a affects the estimates of this integral?* This would be similar to the influence of deficiency of $a \in \Gamma$ on the lengths of Γ-lines (see Theorem 5). It turns out that the previous results of this chapter imply some sharp estimates for the considered integral and also imply a conclusion on its dependence on a. Namely, we along with Sukiasian prove the following results.

Theorem 1. *Let $w(z) \in M_R$ be any function, let $a \in \mathbb{C}$ and let $c = $ const > 1. Then there exists a constant $K(c)$ such that for $r > r_0$*

$$\int_0^1 \frac{L(r, \Gamma(R, a))}{R} dR \le K(c) r T(cr). \tag{4.3.1}$$

For the proof of (4.3.1) it is sufficient to apply the inequality (2.4.1) of Section 2.4 to the equality

$$\int_0^1 \frac{L(r, \Gamma(R, a))}{R} dR = \int\int_{\{z : |z| < r, \, |w(z) - a| < 1\}} \left| \frac{w'(z)}{w(z) - a} \right| d\sigma \tag{4.3.2}$$

which is a consequence of the main identity of Section 1.1.3.

Our next result indicates the sharpness of the estimate (4.3.1) and the dependence of the considered integral mean on the choice of the center a of concentric circles.

Theorem 2. *Let $w(z) \in M_R$ and the geometric deficiency of an $a \in \mathbb{C}$ be greater than zero, i.e. $\bar{\delta}(a) > 0$ (see Section 2.2). Then for $r_0 < r < R$*

$$\int_0^1 \frac{L(r, \Gamma(R, a))}{R} dR \geq (\bar{\delta}(a) - \varepsilon) \int_0^r A(t)dt - h(a) \int_0^r L(t)dt. \quad (4.3.3)$$

For the proof observe that by the definition of geometric deficiencies and Lemmas 1 and 2 of Section 2.2 for $r > r_0$

$$
\begin{aligned}
(\bar{\delta}(a) - \varepsilon) \int_0^r A(t)dt \ &\leq \ \int_0^r \nu(t, a)dt + O(1) \\
&\leq \ \int_0^r \int_{\Delta(r,a)} \left| \frac{\partial}{\partial \varphi} \arg(w(z) - a) \right| d\varphi dt \\
&\quad + h(a) \int_0^r L(t)dt + O(1) \\
&\leq \ \int \int_{\{z:|z|<r,\, |w(z)-a|<1\}} \left| \frac{w'(z)}{w(z) - a} \right| d\sigma \\
&\quad + h(a) \int_0^r L(t)dt + O(1).
\end{aligned}
$$

Hence using (4.3.2) we come to (4.3.3).

The estimate (4.3.3) shows that the integral mean we study is large, provided that the value $a \in \mathbb{C}$ is "bad", i.e. its geometric deficiency is positive, $\bar{\delta}(a) > 0$. Our next theorem shows that conversely for "good" values $a \in \mathbb{C}$ the integral mean in question is "small".

Theorem 3. *Let $w(z) \in M_R$ be any function such that the complete Riemann surface of w^{-1} is not ramified over the disk $\{w : |w - a| < 1\}$.[10] Then*

$$\int_0^1 \frac{L(r, \Gamma(R, a))}{R} dR \leq K(a) r A^{1/2}(r), \quad r < R \quad (4.3.4)$$

for a constant $K(a) < \infty$ depending only on a.

Proof. We use the notations in the proof of Lemma 1 of Section 2.3 and as domains D and D' we take the disks $\{w : |w - a| < 1/2\}$ and $\{w : |w - a| < 1\}$, respectively.

---

[10]Recall that under these conditions we have $\delta(a) = 0$ due to the Kakutani–Tumura–Selberg theorem (see Section 3.1).

By the conditions of our theorem the branch $z_k(w)$ of the function w^{-1}, defined in the disk $w(E'_k(r))$ on the Riemann surface, gives a one-to-one correspondence of this disk to the domain $E'_k(r)$. Therefore we can apply the Koebe inequality (see (2.3.12) of Section 2.3) to the function $z_k(w)$, $w \in w(E'_k(r))$. Hence for $|w - a| \leq 1/2$, $w \in w(E'_k(r))$, we obtain

$$\left| \frac{z'_k(w)}{z'_k(a)} \right| \leq 12.$$

On the other hand, according to the Koebe inequality (2.3.11) of Section 2.3, for $|w - a| = 1/2$, $w \in w(E'_k(r))$.

$$|z'_k(a)| \leq \frac{9}{2}|z_k(w) - z_k(a)| \leq \frac{9}{2}d(E_k(r)).$$

Consequently, for $|w - a| \leq 1/2$. $w \in w(E'_k(r))$.

$$|z'_k(w)| \leq 54 E_k(r).$$

Hence, denoting $\Delta(R) = \{w : w \in w(E'_k(r)), |w - a| = R, z_k(w) \in D(r)\}$ and making use of the above estimates and the main identity of Section 1.1 we get

$$\int_0^1 \frac{L(r, \Gamma(R, a))}{R} dR$$

$$= \int_0^{1/2} \frac{L(r, \Gamma(R, a))}{R} dR + \int_{1/2}^1 \frac{L(r, \Gamma(R, a))}{R} dR$$

$$= \sum_{k=1}^{\Phi(r)} \int_0^{1/2} \frac{1}{R} \int_{\Delta(R)} |z'_k(w)| ds dR$$

$$+ \int\int_{\{z:|z|<r, 1/2<|w-a|<1\}} \frac{|w'(z)|}{|w(z) - a|} d\sigma$$

$$\leq 54\pi \sum_{k=1}^{\Phi(r)} d(E_k(r)) + 2(1 + (1 + |a|^2)) \int\int_{\{z:|z|<r\}} \frac{|w'(z)|}{1 + |w(z)|^2} d\sigma.$$

Applying Lemma 2 of Section 2.3 and Cauchy–Buniakovski inequality to the last inequality completes the proof.

4.4. Distribution of Γ-lines and Value Distribution of Modules and Real Parts of Meromorphic Functions

4.4.1. Here we consider the results of this chapter from the point of view of the approaches mentioned in Section 1.5.2, i.e. from the point of view of "Value Distribution" of the classes $\{|w(z)|\}$, $\{\operatorname{Re} w(z)\}$ and $\{\operatorname{Im} w(z)\}$, where $w(z)$ is a meromorphic function. As we already mentioned in Section 1.5.2, one can construct a "Value Distribution Theory" for these classes by investigating the solutions of equations $|w(z)| = R$ and $\operatorname{Re} w(z) = A$ for different $R \in \mathbb{R}^+$ and $A \in \mathbb{R}$ or for collections of R and A. This is similar to the investigation of solutions of the equality $w(z) = a \in \overline{\mathbb{C}}$ (i.e. of a-points) in Nevanlinna and Ahlfors' Theories.

Rewriting Theorem 2 of Section 4.1, for $\Gamma = \Gamma(R)$ or a collection $\Gamma_1 = \Gamma(R_1), \ldots, \Gamma_q = \Gamma(R_q)$, we immediately obtain the discussed "Value Distribution Theory" for the solutions $|w(z)| = R$, where the main assertions are similar to the classical conclusions of the Nevanlinna Theory (his Second Fundamental Theorem, Deficiency Relation, description of the set of Valiron deficiency values). The classes $\{|w(z)|\}$ and $\{\ln|w(z)|\}$ are very significant subclasses of the class of subharmonic functions. It is known that the main statements of the Nevanlinna Theory are extended to the class of subharmonic functions (see, for example, the monograph of Hayman and Kennedy [1]). One of the main concepts of this extension, namely a magnitude similar to the number of a-points, is introduced in such a way that in the particular case of subharmonic functions of the form $|w(z)|$ it becomes the usual number of a-points of $w(z)$.

In contrast to this, the results of Section 1.5.2 enable us to offer a new general approach, where the analogs of the a-points are the level sets of the equation $|w(z)| = A$. Namely, similar to the results of Section 4.1 for $\nu(x, y) = |w(z)|$, we construct an "alternative Value Distribution Theory".

4.4.2. The situation is somewhat different for the class $\{\operatorname{Re} w(z)\}$ of real parts of meromorphic functions. As the analogs of a-points, here we have to consider the solutions of the equation $\operatorname{Re} w(z) = A$ which is the same as the pre-images of straight lines $\gamma(A) = \{w : \operatorname{Re} w = A\}$ (i.e. $\gamma(A)$-lines, which are the level sets $\operatorname{Re} w(z) = A$ of the function $\operatorname{Re} w(z)$). Theorem 1 of Section 4.1 gives a sharp estimate of $L(r, \gamma(A))$ for a definite value A. But the main Theorem 2 in essence is far from giving an answer for collections $\Gamma_1 = \gamma(A_1), \ldots, \Gamma_q = \gamma(A_q)$. Indeed, if $A_1, \ldots, A_q$ is a collection of pairwise different numbers, then for the function e^z we have

$$\sum_{\nu=1}^{q} L(r, \gamma(A_\nu), e^z) \geq qK_1 rA(r), \quad r > r_0.$$

Hence it follows that some additional restrictions are needed for validity of the inequality (4.1.2) of Theorem 2 for collections of straight lines (and therefore also for collections of infinitely many straight lines). Such additional conditions are given by the following.

Theorem 1. *Let* $w(z) \in M$ *be any function such that the Riemann surface of* w^{-1} *is not ramified in a neighborhood of* ∞. *Further, let* Γ_ν^*, $\nu = 1, 2, \ldots, q$, *be a collection of disjoint, unbounded smooth Jordan curves with disjoint* $\nu(\Gamma_\nu^*) < \infty$, $\nu = 1, 2, \ldots, q$.
Then there exist some absolute constants $K < \infty$ *and* $c \in (0, 1)$ *such that*

$$\sum_{\nu=1}^{q} L(r. \Gamma_\nu^*) \leq KrA(r) \qquad (4.4.1)$$

for any $r \in E$, *where* E *is a set of lower logarithmic density* $\mathbb{C}$.

This result easily follows from Theorems 2 and 4 of Section 4.1. Indeed, the conditions of Theorem 1 imply the existence of a constant τ such that the considered Riemann surface is not ramified over the exterior of the disk $D(\tau) = \{w : |w| < \tau\}$. For a fixed $\tau_0 > 2\tau$ we split each of our curves Γ_ν^* into $\Gamma_{\nu,1}^*(\tau_0) = \Gamma_\nu^* \bigcap \{w : |w| \leq \tau_0\}$ and $\Gamma_{\nu,2}^*(\tau_0) = \Gamma_\nu^* \bigcap \{w : |w| > \tau_0\}$. As $\nu(\Gamma) < \infty$, one can be convinced that if τ_0 is large enough, then each set $\Gamma_{\nu,1}^*(\tau_0)$ consists of only one curve and each set $\Gamma_{\nu,2}^*(\tau_0)$ consists of not more than two unbounded curves.[11] Now we get the inequality (4.4.1) by means of applying Theorem 2 of Section 4.1 to the curves $\Gamma_{\nu,1}^*(\tau_0)$ and Theorem 4 of Section 4.1 to the sum of curves $\bigcup_\nu \Gamma_{\nu,2}^*(\tau_0)$ and then by summing up, the estimates are obtained.

Particularly, when $\Gamma_\nu^* = \gamma(A_\nu)$, $\nu = 1, 2, \ldots, q$, and A_ν are pairwise different, the inequality (4.4.1) can be considered as an analog of the Second Fundamental Theorem for the distribution of real parts of the considered class of meromorphic functions.

---

[11] Under the above conditions the set Γ_ν^* consists of one curve if Γ^* is an open Jordan curve with endpoints $z \neq \infty$ and $z = \infty$ and Γ_ν^* consists of two curves if Γ^* is a closed Jordan curve involving the point at infinity.

4.5. The Number of Γ-lines Crossing Rings

4.5.1. Let $J(r, \Gamma, k)$, $0 < k < 1$, be the number of those connected components of the set $w^{-1}(\Gamma) \bigcap \{z : |z| < r\}$ which cross both circles of the ring $\{z : kr \leq |z| \leq r\}$. This magnitude somewhat characterizes the asymptotic behavior of the meromorphic function w since any asymptotic line with w-image over Γ cross both these circles for all $r > r_0$.

Estimates of $J(r, \Gamma, k)$ immediately follow from the above estimates of $L(r, \Gamma)$ since obviously

$$J(r, \Gamma, k) \leq J\left(\frac{1+k}{2}r, \Gamma, \frac{2k}{1+k}\right) \leq \frac{2(L(r, \Gamma) - L(kr, \Gamma))}{(1-k)r}. \qquad (4.5.1)$$

Application of this inequality to the estimates of L established in Section 4.1 leads to the following two theorems.

Theorem 1. *Under the conditions of Theorems 1 and 2 of Section 4.1 the inequalities (4.1.1) and (4.4.1′) of Theorem 1 can be replaced by*

$$J(r, \Gamma, k) \leq K(k)A(r), \quad r \in E \qquad (4.5.2)$$

and

$$J(r, \Gamma, k) \leq K(k)K(\Gamma)T(r), \quad r > r_0 \qquad (4.5.3)$$

and the inequality (4.1.2) of Theorem 2 can be replaced by

$$\sum_{\nu=1}^{q} J(r, \Gamma_\nu, k) \leq K(k)A(r), \quad r \in E. \qquad (4.5.4)$$

Here $K(k) = \text{const} < \infty$.

The inequalities (4.5.2) and (4.5.4) immediately follow from (4.1.1) and (4.1.2) of Section 4.1 after applying (4.5.1). For proving (4.5.3) it suffices to take $(1 + k)r/2$ instead of r and $2/(1 + k)$ instead of c in the inequality (4.1.1′) of Section 4.1. Then we get

$$L\left(\frac{1+k}{2}r, \Gamma\right) \leq K(k)K(\Gamma)rT(r),$$

whence applying (4.5.1) we obtain (4.5.3).

The inequality (4.5.4) obviously implies an analog of Corollary 1 of Section 4.1, where $\liminf_{r\to\infty} J(r, \Gamma, k)/A(r)$ substitutes $\liminf_{r\to\infty} L(r, \Gamma)/(rA(r))$. Theorem 3 of Section 4.1 implies the following result for similar upper bounds.

Theorem 2. *The equality* (4.1.7) *of Theorem 3 from Section* 4.1 *can be replaced by*

$$\limsup_{r\to\infty} \frac{J(r, \Gamma(R), k)\varphi(r)}{A^{1/2}(r)} = 0.$$

4.6. Distribution of Gelfond points

4.6.1. The following result was proved by Gelfond [1] in 1934: if $w(z)$ is an entire function of the order ρ and Γ is the real axis, then

$$\limsup_{r\to\infty} \frac{\ln \Phi(r, \Gamma)}{\ln r} = \rho, \tag{4.6.1}$$

where $\Phi(r, \Gamma) = \Phi(r, \Gamma, w)$ is the number of those points z on the circle $|z| = r$ whose images under w lie on Γ; let us call them *Gelfond points*.

Much later, this result was reproved by Wilf [1], and also by Hellerstein and Korevaar [1]. After this, the order of $\Phi(r, \Gamma)$ was evaluated in a series of papers. Perhaps, the most complete results in this field are given by Hellerstein [1] and by Miles and Townsend [1.2]. Particularly, Hellerstein [1] proved that if $w(z)$ is an entire function with the lower order λ and Γ is an unbounded algebraic curve, then along with (4.6.1)

$$\liminf_{r\to\infty} \frac{\ln \Phi(r, \Gamma)}{\ln r} = \lambda. \tag{4.6.2}$$

In the same paper it was proved that (4.6.2) is valid if Γ is a bounded algebraic curve with closure containing a Nevanlinna deficiency value of $w(z)$.

The above mentioned researches of Miles and Townsend contain the following results. If $w(z)$ is an entire function and Γ is the circle $|w| = 1$, then

$$\limsup_{r\to\infty} \frac{\ln \Phi(r, \Gamma)}{\ln r} \leq \rho. \tag{4.6.3}$$

This answers a question of Korevaar in Hayman's survey [4] of unsolved problems.

In the same place it is shown that there exists a meromorphic function in $|z| < \infty$ of arbitrary slow growth for which the function $\Phi(r, \Gamma)$ increases more rapidly on $r_n \to \infty$ than any previously given function, where Γ is a straight line. Hence the relation (4.6.1) is not true for meromorphic functions. Nonetheless, the next theorem establishes some similar results for the lower orders of functions meromorphic in $|z| < \infty$ and for a significantly broader class of curves Γ.

Theorem 1. *Let $w(z) \in M$ and let Γ be a smooth Jordan curve with $\nu(\Gamma) < \infty$. Then*

$$\liminf_{r \to \infty} \frac{\ln \Phi(r, \Gamma)}{\ln r} \leq \lambda. \tag{4.6.4}$$

Besides, if $\Phi^*(r, \Gamma) := \int_0^r \Phi(t, \Gamma) dt$, then

$$\limsup_{r \to \infty} \frac{\ln \Phi^*(r, \Gamma)}{\ln r} \leq \rho + 1 \tag{4.6.5}$$

and

$$\liminf_{r \to \infty} \frac{\ln \Phi^*(r, \Gamma)}{\ln r} \leq \lambda + 1. \tag{4.6.6}$$

Evidently, the relation (4.6.4) is a very rough estimate for $\Phi(r, \Gamma)$ (since the lower order of the function $\Phi(r, \Gamma)$ is estimated instead of the function). A direct estimate of $\Phi(r, \Gamma)$ was obtained by Miles and Townsend [1, 2] for the case when f is a meromorphic function in $|z| < \infty$ and Γ is a straight line. They were able to prove that

$$\Phi(r, \Gamma) = O(\alpha(r)) T(Br, f) \quad \text{as} \quad r \to \infty, \quad r \notin E, \tag{4.6.7}$$

where $\alpha(r)$ is an increasing, unbounded function on $[0, \infty)$, $B = \text{const} > 1$, and $E \subset [0, \infty)$ is a set of zero logarithmic density.

The next theorem establishes some sharp estimates of $\Phi(r, \Gamma)$ and of sums $\sum_{\nu=1}^q \Phi(r, \Gamma_\nu)$ for a broad class of curves Γ. It leads to some conclusion that is similar to the Deficiency Relation, but related to "distribution of Gelfond magnitudes".

Theorem 2. *Let $w(z) \in M$ be a function with a lower order $\lambda < \infty$ and let Γ_ν, $\nu = 1, 2, \ldots, q$, be disjoint, smooth Jordan curves with $\nu(\Gamma_\nu) < \infty$, such that $\Gamma_1, \ldots, \Gamma_{q-1}$ are bounded and Γ_q is unbounded.*

Then for an unbounded set of r

$$\sum_{\nu=1}^{q} \Phi(r, \Gamma_\nu) \le K(\lambda) T(r),\qquad (4.6.8)$$

where $K(\lambda) < \infty$ is a constant depending solely on λ.

Remark. By the relation (1.5.4) of Section 1.5.5

$$\Phi^*(r, \Gamma) \le L(r, \Gamma).\qquad (4.6.9)$$

Therefore, by Theorem 2 of Section 4.1 the inequality (4.1.2) in Theorem 2 can be replaced by

$$\sum_{\nu=1}^{q} \int_0^r \Phi(t, \Gamma) dt \le K r A(r),\qquad (4.6.10)$$

where K already is an absolute constant.

Theorem 2 implies the following similarity of the deficiency relation.

Corollary 1. *If $w(z) \in M$, $\lambda < \infty$ and $M(\Gamma)$ is as defined in Section 4.1, then the following relations are valid for any curve $\Gamma \in M(\Gamma)$, except not more than a countable set of curves $\Gamma_\nu \in M(\Gamma)$:*

$$\liminf_{r \to \infty} \frac{\Phi(r, \Gamma)}{T(r)} = 0,\qquad (4.6.11)$$

$$\sum_{\nu=1}^{q} \liminf_{r \to \infty} \frac{\Phi(r, \Gamma_\nu)}{T(r)} = 0.\qquad (4.6.12)$$

4.6.2. Proofs of Theorem 1 and 2. For proving Theorem 1 suppose that the lower limit in (4.6.4) is attained over a sequence $r_n \to \infty$. By the mean value theorem for any term r_n there exists $r'_n \in [r_n/2, r_n]$ for which

$$\Phi(r'_n, \Gamma) = \frac{2}{r_n} \int_{r_n/2}^{r_n} \Phi(t, \Gamma) dt.$$

Consequently

$$\liminf_{r \to \infty} \frac{\ln \Phi(r, \Gamma)}{\ln r} \le \liminf_{r'_n \to \infty} \frac{\ln \Phi(r'_n, \Gamma)}{\ln r'_n} \le \liminf_{r \to \infty} \frac{\ln \Phi^*(r_n, \Gamma)}{\ln r_n} - 1,$$

and hence (4.6.4) follows from the inequality (4.6.6). For proving (4.6.5) and (4.6.6) note that by the principle of length and tangent variation and (4.6.9)

$$\Phi^*(r, \Gamma) \leq K(\Gamma)[V(D(r)) + l(D(r))],$$

where by the inequality (1.2.9) of Section 1.2.2 and Lemma 1 of Section 2.4

$$V(D(r)) \leq 2 \iint_{D(r)} \left| \frac{w''(z)}{w'(z)} \right| d\sigma \leq K(c)rT(cr), \quad c = \text{const} > 0. \quad (4.6.13)$$

On the other hand, obviously $l(D(r)) = 2\pi r$. Therefore,

$$\Phi^*(r, \Gamma) \leq K(\Gamma, c)rT(cr), \quad K(\Gamma, c) = \text{const} < \infty$$

for $r > r_0$, whence (4.6.5) and (4.6.6) hold, and the proof of Theorem 1 is complete.

For proving Theorem 2 we consider the parts of the curve Γ_q lying in $D(\tau)$ and its exterior. As $\nu(\Gamma) < \infty$, we can choose τ so that the part of Γ_q in $D(\tau)$ is a single curve Γ_q^* and the other part of Γ_q, which lies in the exterior of $D(\tau)$, consists of not more than two curves $\Gamma_q(1)$ and $\Gamma_q(2)$ with $\nu(\Gamma_q(1)) < \pi/4$ and $\nu(\Gamma_q(2)) < \pi/4$.

By the principle of length and tangent variation written in the form of the inequality (1.2.15) of Section 1.2.3 and (4.6.9), (4.6.13),

$$\Phi^*(r, \Gamma_q(\mu)) \leq \frac{4\sqrt{2}}{\pi} V(D(r)) + \sqrt{2}l(D(r)) \leq K(c)rT(cr), \quad \mu = 1, 2.$$

Further, using (4.6.9) and (4.6.13) and also the Second Fundamental Theorem of Section 1.3.2 and Lemma 1 of Section 2.4 we get

$$\begin{aligned}
\sum_{\nu=1}^{q-1} \Phi^*(r, \Gamma_\nu) + \Phi^*(r, \Gamma_q^*) &\leq \sum_{\nu=1}^{q} L(r, \Gamma_\nu) + L(r, \Gamma_q^*) \\
&\leq \frac{\sqrt{28}}{\pi} V(D(r)) + h(\Gamma_1, \ldots, \Gamma_{q-1}, \Gamma_q^*) \\
&\quad \times \int_0^r L(t)dt + \sqrt{2}l(D(r)) \\
&\leq \frac{\sqrt{28}}{\pi} K(c)rT(cr) + o(rT(cr)), \quad r \to \infty.
\end{aligned}$$

$$(4.6.14)$$

Summing up previous inequalities we obtain

$$\sum_{\nu=1}^{q} \Phi^*(r, \Gamma_\nu) := \sum_{\nu=1}^{q-1} \Phi^*(r. \Gamma_\nu) + \Phi^*(r, \Gamma_q^*) + \Phi^*(r, \Gamma_q(1))$$
$$+ \Phi^*(r. \Gamma_q(2)) \le K_0 K(c) r T(cr), \quad r > r_0,$$

where K_0 is an absolute constant. Taking into account that for any r there exists $\rho_r \in [r/2, r]$ such that

$$\Phi(\rho_r, \Gamma) = \frac{2}{r} \int_{r/2}^{r} \Phi(t, \Gamma) dt \le \frac{2}{r} \Phi^*(r, \Gamma),$$

from the last inequality we conclude that for $r > r_1$

$$\sum_{\nu=1}^{q} \Phi(\rho_r, \Gamma_\nu) \le 2 K_0 K(c) T(cr).$$

On the other hand, by Lemma 6 of Section 2.4.3 there exists a sequence $r_n \to \infty$ such that $T(cr_n) \le c^{2(\lambda+1)} T(r_n/c)$. Consequently, for a fixed c, $r_n \to \infty$ and $r_n > r_2$

$$\sum_{\nu=1}^{q} \Phi(\rho_{r_n}, \Gamma_\nu) \le K(\lambda) T(r_n/2) \le K(\lambda) T(\rho_{r_n}),$$

i.e. the desired inequality (4.6.8) is obtained and hence Theorem 2 is proved.

4.7. Nevanlinna's Dream-description of Transcendental Ramification of Riemann Surfaces

4.7.1. History and main ideas.
One of the most remarkable features of the Value Distribution Theory is its rich geometric essence providing natural interpretation of the main concepts and relation of the theory on the Riemann surfaces. This interpretation is based on heuristic remarks of Nevanlinna ([1] Chapter XII, Sections 1.2 and [2], Section 15) connecting his Second Fundamental Theorem and Riemann's formula on the total ramification of the Riemann surfaces of rational functions.[12] Here the guiding thrust toward

---

[12]Later Ahlfors [3] gave a beautiful confirmation of this connection. His main result is an immediate generalization of the Hurwitz inequality (contained in the Hurwitz formula, which in turn, is a generalization of the Riemann formula). See Nevanlinna [1, 2, 3]; Stoilov [1, 2].

geometric considerations was so strong that Nevanlinna ([2], p. 352) proposed that the final object in the Value Distribution Theory would have to be a precise knowledge of the corresponding Riemann surface. Of course, such a knowledge must contain, in the first place, information about singularities of the Riemann surface of the inversions w^{-1} of meromorphic functions w. Such an information is usually understood as a description of the "cardinality" of the algebraic and transcendental points of the Riemann Surface that lie over the previously given values in the w-plane.[13]

For the particular case when the Riemann surfaces under consideration are from the class F_q (the class of Riemann surfaces whose sets of singularities project onto finitely many points), the mentioned cardinality can be determined by means of a simple and beautiful construction (see Nevanlinna [1], Chapter XII, Sections 1, 2; [2], Section 4). In this construction the surface F_q is broken up into sheets on which the numbers of algebraic and logarithmic[14] singularities connected with each of these sheets is calculated. Then F_q is exhausted by "bundles" of these sheets and the number of singularities corresponding to these bundles is calculated.

However, it was repeatedly noted in the literature that such an effective partitioning of Riemann surfaces is scarcely possible. In the general case, the exhaustion of the complete Riemann surface of w^{-1} is usually accomplished by means of the surfaces $F_r = \{w(z) : |z| \leq r\}$ by letting $r \to \infty$. After the use of certain relations for the disk $|z| \leq r$ (which is the same as for F_r) the passage $r \to \infty$ leads to some characteristics of algebraic and transcendental ramification. As such a relation Nevanlinna offered his Second Fundamental Theorem. The following is his statement (see [2]) on the description of singularities of Riemann surfaces by means of relations in value distribution theory: *One can presuppose that to any deficiency value always corresponds a transcendental singularity in the general case or, what is reduced to the same by Iversen's theorem, any value having a positive defect always is an asymptotic value for the considered meromorphic function.*

---

[13]The transcendental points of the Riemann surface of the function w^{-1} are defined to be its non-algebraic critical points. By the latter we mean a critical point of a branch of the Riemann surface, i.e. a point through which this branch cannot be continued in a single-valued way or cannot be continued at all.

[14]In the considered case every singularity (i.e. critical point) is isolated, i.e. in its sufficiently small neighborhood the corresponding branch can be continued without a restriction and without meeting other singularities. One can easily show that only the algebraic and logarithmic points are isolated critical points.

Although this conjecture of Nevanlinna has so far been proved only under special assumptions, it is nevertheless useful, if only to give a correct intuitive idea about the interrelations of the problems considered here and to understand the deficient values as quantities indicating the presence of transcendental singularities of the inverse function. According to such an understanding the Second Fundamental Theorem stating that

$$\sum_{a \in \overline{\mathbb{C}}} \Theta(a) = \sum_{a \in \overline{\mathbb{C}}} \left(1 - \limsup_{r \to \infty} \frac{\overline{N}(r, a)}{T(r)} \right) \leq 2$$

(i.e. the sum of the deficient quantities Θ cannot exceed 2) can be regarded as an assertion about the ramification of the Riemann surface of a meromorphic function. The index $\Theta(a)$ determines the branching degree of the Riemann surface over the point a. On the other hand, it breaks up into two components $\vartheta(a)$ and $\delta(a)$, where $\vartheta(a)$ is the branching index characterizing the algebraic ramification point lying over $w = a$ and $\delta(a)$ is the deficiency characterizing the branching points of infinite order, i.e. transcendental points (see Nevanlinna [2], p. 379).

Indeed, the Value Distribution Theory contains assertions of remarkable generality and precision, about the algebraic singularities of Riemann surfaces:

$$\liminf_{r \to \infty} \frac{N_1(r)}{T(r)} \leq 2.$$

As to the proposal to regard the deficiency as a characteristic of transcendental ramification, it is based on the following observation. The fact that a is a deficiency value (i.e. $\delta(a) > 0$) or, which is the same as

$$\limsup_{r \to \infty} \frac{N(r, a)}{T(r)} < 1$$

means that the value a is taken by $w(z)$ relatively rarely in any disk $|z| \leq r$. Therefore it seems natural to suppose that the origin of the deficient value is due to the fact that a is rarely taken on the Riemann surface of w^{-1}, i.e. the set of transcendental singular points in the Riemann surface of w^{-1}, which lie over a has a sufficiently high cardinality. Moreover, the value of deficiency $\delta(a)$ would have to be as larger and stronger as the transcendental ramification of the Riemann surface over a.

However, later it became evident that the concept of the deficiency is not directly connected with the transcendental ramification of Riemann surfaces, since examples of non-asymptotic deficiency values have been constructed for different classes of functions[15] (see the survey by Goldberg [2], Section 15). Moreover, the deficiency value of some value a can be exactly equal to 1 without the value a being asymptotic.

The mentioned examples show that not only the transcendental ramification but also some other factors influence the magnitude $m(r, a)$. Therefore it is natural to try to isolate the portion of the deficiency adequate to transcendental ramification. Below we establish such an adequacy for particular case whose analysis seems to reveal the geometric mechanism for the origination of deficiency and various factors' influence on it.

Consider the case when all the singularities in the part of the Riemann surface of w^{-1} lying over some disk $|w-a| < \rho$ are projected to the point a. As we mentioned in the beginning of this section, in this case the transcendental points over a are logarithmic. We draw straight line cuts from points of the Riemann surface lying over a to the points lying over $a+\rho$. These cuts divide the Riemann surface over $|w - a| < \rho$ into infinitely many sheets which are discs cut from a to $a+\rho$. Now suppose that the boundary ∂F_r of the surface $F_r = \{w(z) : |z| \le r\}$ has common points with some of these cut disks and split the latter in collections $\{G_l\}$, $\{G_a\}$ and $\{G_u\}$, where the point a is a logarithmic, algebraic branching point and usual (ordinary) point. Further, we suppose that for one of these collections (say G_l) there are k connected components τ_i, $i = 1, 2, \ldots, k$, of the set $\partial F_r \cap G_l$, each visible from a in an angle of opening 2π. We assign a number J_i equal to $+1$ or -1 to every such component, dependent on whether the growth of $\arg(w(z) - a)^{-1}$ on τ_i is 2π or -2π. Then we set

$$\nu(r, a)_l = \sum_{\{G_l\}} \sum_{i=1}^{k} J_i, \quad \nu(r, a)_a = \sum_{\{G_a\}} \sum_{i=1}^{k} J_i, \quad \nu(r, a)_u = \sum_{\{G_u\}} \sum_{i=1}^{k} J_i.$$

The following relations are either geometrically clear or can be established by means of the arguments proving Lemma 2 of Section 2.2.3.

(i) $\left| \sum_{i=1}^{k} J_i \right| \le 1$ for G_l, G_a and G_u, and for G_l also $\sum_{i=1}^{k} J_i \ge 0$.

(ii) $\nu(r, a)_o = hL(r)$, where $|h| < h(\rho) = \text{const} < \infty$.

(iii) $\nu(r, a)_l + \nu(r, a)_a + \nu(r, a)_u = \nu(r, a) + hL(r)$, where $|h| < h(\rho) = \text{const} < \infty$.

---

[15] According to Iversen's theorem, if a is not an asymptotic value for $w(z)$, then the Riemann surface of w^{-1} does not have transcendental singularities over a.

From (ii) and Theorem 1 of Section 2.2.2 we get

$$n(r,a) + \nu(r,a)_l + \nu(r,a)_a + \nu(r,a)_u = A(r) + hL(r),$$

where $|h| < h(\rho) = \text{const} < \infty$. It is evident that here $\{\nu(r,a)_l + \nu(r,a)_a\}$ takes the same role that $m(r,a)$ takes in the First Fundamental Theorem of Nevanlinna. However, it is remarkable that here we have separated the magnitude $\nu(r,a)_l$ corresponding to logarithmic singularities. Therefore, it is natural to regard the quantity

$$\liminf_{r \to \infty} \frac{\nu(r,a)_l}{A(r)}$$

as a characteristic of the transcendental ramification over a. On the other hand, the "geometric deficiency"

$$\liminf_{r \to \infty} \frac{\nu(r,a)}{A(r)}$$

is influenced also by $\nu(r,a)_a$. For clarifying this influence we consider also the pre-images of the above mentioned cuts partitioning the Riemann surface over $|w - a| < \rho$.

If $\nu(r,a)_a$ is large, then by (i) there must exist pre-images of algebraic points of large multiplicity, which lie outside $|z| \leq r$, and many pre-images of the cuts connected with these algebraic points have to intersect $|z| = r$. This has to do with how the quantity $\nu(r,a)_l$ is generated; in the last case we are dealing with a logarithmic point whose multiplicity, roughly speaking, is infinite, and the number of intersections of $|z| = r$ with the pre-images of the corresponding cuts increases as r increases. In addition, the algebraic points with a large sum of multiplicities take the same role as the logarithmic points in the sense of their contribution to $\nu(r,a)$. Thus, by extracting $\nu(r,a)_l$ and $\nu(r,a)_a$ we distinguish two cases when $|z| = r$ crosses the unbounded pre-images of the cuts and when it intersects the bounded pre-images of the cuts; this obviously depends on whether the corresponding cut is connected with a logarithmic or algebraic point. In essence this is the basic idea of our approach to the description of transcendental singularities in the general case.

4.7.2. Theorems on ramification of Riemann surfaces over curves: a deficiency relation.

Let $\mathfrak{M}_\Gamma$ be the set of all points of the complete Riemann surface of the function w^{-1}, together with the boundary points projected to Γ. Remove from $\mathfrak{M}_\Gamma$ all its algebraic and transcendental points. The remaining set contains a collection of open arcs $\mathfrak{m}_i(\Gamma)$ from the Riemann surface, the endpoints of which are algebraic, transcendental or ordinary points (in the last case these points lie over the endpoints of Γ). Let $\mathfrak{M}_1(\Gamma) = \{\mathfrak{m}_i(\Gamma)\}$ be the set of these arcs (here i is the number of an arc) and let $\mathfrak{M}_2(\Gamma) \subset \mathfrak{M}_1(\Gamma)$ be the subset of those arcs $\mathfrak{m}_\mu^0(\Gamma)$ which have at least one algebraic or transcendental endpoint. Let $\overline{C}(r, \Gamma)$ be the number of those arcs in $\mathfrak{M}_1(\Gamma)$, the pre-images of which have common points with the disk $|z| \leq r$, and let $C(r, \Gamma)$ be the number of the same type of arcs in $\mathfrak{M}_2(\Gamma)$.

The following theorems hold, see Barsegian [3]–[9].

Theorem 1. *Let $w(z) \in M$ be any function with $\lambda < \infty$, and let Γ be a smooth Jordan curve with $\nu(\Gamma) < \infty$. Then there exists a constant $K(\lambda) < \infty$ depending solely on λ, such that*

$$\overline{C}(r, \Gamma) < K(\lambda)T(r) \tag{4.7.1}$$

over a sequence $r = r_n \to \infty$.

Theorem 2. *Let $w(z) \in M$ be any function with $\lambda < \infty$, and let Γ_ν, $\nu = 1, 2, \ldots, q$, be any disjoint, smooth Jordan curve with $\nu(\Gamma) < \infty$, such that $\Gamma_1, \ldots, \Gamma_{q-1}$ are bounded and Γ_q is unbounded.*

Then there exists a constant $K(\lambda) < \infty$ depending solely on λ, such that

$$\sum_{\nu=1}^{q} C(r, \Gamma_\nu) < K(\lambda)T(r) \tag{4.7.2}$$

over a sequence $r = r_n \to \infty$.

We remark that the pre-image of any arc $\mathfrak{m}_\mu^0(\Gamma) \in \mathfrak{M}_2(\Gamma)$ intersects $|z| \leq r$ for $r > r_0$. Therefore, the quantity $C(r, \Gamma)$ gets only unit increment for these r, i.e. $C(r, \Gamma)$ acts as a "counter" of the arcs of $\mathfrak{M}_2(\Gamma)$ and by the same token a "counter" of the singularities of the Riemann surface connected with these arcs. This simple remark shows that $C(r, \Gamma)$ can be considered as a quantity which in a certain indirect way characterizes the "cardinality" of the algebraic and transcendental singular points lying over Γ.

Now we shall show a way of extracting the contributions describing the algebraic and transcendental ramifications of Riemann surfaces. We consider singularities w_0 , which are projected into Γ and are not its endpoints. We denote the number of arcs of the form $\mathfrak{m}_\mu^0(\Gamma)$ intersecting with the disk $|z| < r$ by $\mathfrak{m}_\mu^0(r, \Gamma)$. Further, we assume that $\vartheta(r, \Gamma)$ is the number of algebraic ramification points which are endpoints of the arcs $\mathfrak{m}_\mu^0(\Gamma)$. Also we assume that $\delta(r, \Gamma)$ is the number of transcendental ramification points that are endpoints of the arcs $\mathfrak{m}_\mu^0(\Gamma)$. Obviously any arc $\mathfrak{m}_\mu^0(\Gamma)$ can increment both $\vartheta(r, \Gamma)$ and $\delta(r, \Gamma)$ by exactly 1. Also it is obvious that

$$\vartheta(r, \Gamma) + \delta(r. \Gamma) \le 2C(r, \Gamma). \tag{4.7.3}$$

Suppose that w_0 is an algebraic ramification point of the multiplicity k. Then on the complete Riemann surface of w^{-1} there necessarily exist exactly $2k$ arcs $\mathfrak{m}_\mu^0(\Gamma)$ having w_0 as an endpoint. Therefore, the contribution of this algebraic ramification point to $\vartheta(r, \Gamma)$ is equal to $2k$ as $r > |w^{-1}(w_0)|$. Consequently, it is natural to regard the quantity

$$\vartheta(\Gamma) := \liminf_{r \to \infty} \frac{\vartheta(r, \Gamma)}{2T(r)}$$

as the total ramification of algebraic ramification points lying over Γ. Similarly, the quantity

$$\delta(\Gamma) := \liminf_{r \to \infty} \frac{\delta(r, \Gamma)}{2T(r)}$$

can be regarded as the total ramification of transcendental ramification points lying over Γ. This point of view is clearly justified by the consideration of the logarithmic ramification points w_0. For the latter, the arcs $\mathfrak{m}_{\mu_1}^0(r, \Gamma)$ and $\mathfrak{m}_{\mu_2}^0(r, \Gamma)$, lying on the same sheet of the complete Riemann surface and both having w_0 as an endpoint, due to w_0 contribute 2 to $\delta(r, \Gamma)$. Therefore the quantities $\delta(r, \Gamma)/2$ on average describe the number of those sheets in the neighborhood of w_0, the pre-images of which have common points with the disk $|z| < r$.

By (4.7.2) and (4.7.3) we come to the following similarity of the deficiency relation.

Corollary 1. *Let $w(z) \in M$ be any function with $\lambda < \infty$ and let $M(\Gamma)$ be that defined in Section 5.1. Then for any $\Gamma \in M(\Gamma)$, with the possible exception of a countable set $\Gamma_\nu \in M(\Gamma)$,*

$$\Theta(\Gamma) + \delta(\Gamma) = 0 \tag{4.7.4}$$

and

$$\sum_{\{\Gamma_\nu\}} \Theta(\Gamma_\nu) + \sum_{\{\Gamma_\nu\}} \delta(\Gamma_\nu) \leq K(\lambda). \qquad (4.7.5)$$

The magnitude $\delta(\Gamma)$ (i.e. the deficiency of values on the curve Γ which is the same as the total transcendental ramification over Γ) can be regarded as an indicator of the transcendental ramification over any point $w_0 \in \Gamma$ and even over any set $w_1, w_2, \ldots \in \Gamma$. Assuming $\delta_0(r, \Gamma, w_0)$ to be the number of arcs $\mathfrak{m}_\mu^0(r, \Gamma)$ having w_0 as an endpoint, we define the "deficiency value" $\delta_0(w_0)$ over w_0 as follows:

$$\delta_0(w_0) = \liminf_{r \to \infty} \frac{\delta_0(r, \Gamma, w_0)}{2T(r)}.$$

Accordingly, we call w_0 a deficient value if $\delta_0(w_0) > 0$. It is clear that w_0 is a deficient value only when Γ is a deficient curve, i.e. $\delta(\Gamma) > 0$. Now (4.7.5) implies

$$\sum_{(a)} \delta_0(a) < K(\lambda). \qquad (4.7.6)$$

4.7.3. A connection with a hypothesis of Nevanlinna. Nevanlinna supposed that the number of the deficient values of an entire function of the order $\rho < \infty$ probably does not exceed $[2\rho]$ (i.e. the entire part of 2ρ). This was contradicted by Arakelian [1] who showed that there exist entire functions of any order $\rho \in (1/2, \infty)$, having infinite sets of deficient values. Nonetheless, if the deficiency of $w_0 \in \Gamma$ is understood as in the previous subsection, then this Nevanlinna's hypothesis turns to be true. By the well-known theorem of Denjou–Carleman–Ahlfors, the number of asymptotic values of an entire function of the order $\rho < \infty$ does not exceed $[2\rho]$. If the asymptotic values are $a_1, a_2, \ldots, a_q$, $q \leq [2\rho]$, then by Iversen's theorem the transcendental points can lie only over $a_1, a_2, \ldots, a_q$. Consequently, the number of deficient values $\delta_0(a)$ in the sense of Section 4.7.2 does not exceed $[2\rho]$.

4.7.4. A comparative analysis of descriptions of transcendental ramifications. It is useful to compare the inequality (4.7.2) with the Second Fundamental Theorem written in the form of the inequality (2.1.33) of Section 2.1.10, since both these inequalities deal with some sets on the Riemann surface, which are projected onto Γ.

Obviously

$$m(r, \Gamma) + n_1(r, \Gamma) \leq C(r, \Gamma),$$

hence it follows that (4.7.2) in Theorem 2 can be replaced by the inequality

$$\sum_{\nu=1}^{q} [m(r, \Gamma_\nu) + n_1(r, \Gamma_\nu)] \leq K(\lambda)T(r),$$

which is a less precise version of the inequality (2.1.34′) of Section 2.1.10. In the mentioned inequality of Ahlfors only the magnitudes $m(r, \Gamma)$ can be related to transcendental points. Therefore, only the contributions of those $\mathfrak{m}_\mu^0(\Gamma) \in \mathfrak{M}_2(\Gamma)$ that have at least one transcendental endpoint are of interest from the point of view of transcendental ramification. We shall denote such an $\mathfrak{m}_\mu^0(\Gamma)$ by $\mathfrak{m}_\mu^*(\Gamma)$. Let the set of all arcs $\mathfrak{m}_\mu^*(\Gamma)$ be somehow partitioned into collections $\mathfrak{m}_k$ of arcs with disjoint projections (where k is the number of a collection). The following remark answers how the magnitudes $m(r, \Gamma)$ and $C(r, \Gamma)$ take the singularities into account.

Remark. If there are n_k arcs $\mathfrak{m}_\mu^*(\Gamma)$ in the collection $\mathfrak{m}_k$, then the contribution of $\mathfrak{m}_k$ to $m(r, \Gamma)$ is less than 1 for any r, while its contribution to $C(r, \Gamma)$ is not less than n_k. If there is a countable set of arcs $\mathfrak{m}_\mu^*(\Gamma) \in \mathfrak{m}_k$, then the contribution of $\mathfrak{m}_k$ to $m(r, \Gamma)$ is less than 1 for any r, while its contribution to $C(r, \Gamma)$ tends to ∞ as $r \to \infty$.

4.7.5. Proof of Theorem 1. Assume that $c > 1$ is a fixed number and denote by $l_\mu(r, \Gamma)$ a curve which has common points with the disk $|z| \leq r$ and is a pre-image of an arc $\mathfrak{m}_\mu(\Gamma) \in \mathfrak{M}_1(\Gamma)$. We break the set of such $l_\mu(r, \Gamma)$ into the following three groups.

(i) $l_\mu(r, \Gamma)$ is completely contained in $|z| \leq cr$ and as an endpoint it has at least one multiple point. The number of such $l_\mu(r, \Gamma)$ we denote by $N_1(r, \Gamma)$.

(ii) $l_\mu(r, \Gamma)$ is completely contained in $|z| \leq cr$ and its endpoints are not multiple points. The number of such $l_\mu(r, \Gamma)$ we denote by $N_2(r, \Gamma)$.

(iii) $l_\mu(r, \Gamma)$ has common points with $\{z : |z| > cr\}$. The number of such $l_\mu(r, \Gamma)$ we denote by $N_3(r, \Gamma)$.

Evidently

$$N_1(r, \Gamma) \leq 2n_1(cr). \tag{4.7.7}$$

where $n_1(cr)$ is the sum of orders of multiple points of the function $w(z)$, which lie in the disk $|z| \leq cr$. Further, any curve of (ii) is a pre-image of a simple island (i.e. island of the multiplicity 1) over Γ for the surface $\widetilde{F}_{cr}$. Consequently, by Ahlfors' theorem

$$N_2(r, \Gamma) \leq A(cr) + h(\Gamma)L(cr), \tag{4.7.8}$$

where $h(\Gamma) = \mathrm{const} < \infty$. At last, as the length of any curve of (iii) is greater than $(c-1)r$,

$$N_3(r, \Gamma) \leq \frac{L(cr, \Gamma) - L(r, \Gamma)}{(c-1)r},$$

and by the inequality (4.6.14) of Section 4.6

$$N_3(r, \Gamma) \leq K(c)T(cr), \quad r > r_0. \tag{4.7.9}$$

It is obvious that $\overline{C}(r, \Gamma) \leq N_1(r, \Gamma) + N_2(r, \Gamma) + N_3(r, \Gamma)$. Therefore by the inequalities (4.7.7)–(4.7.9)

$$\overline{C}(r, \Gamma) \leq 2n_1(cr) + A(cr) + hL(cr) + K(c)T(cr).$$

Hence, taking into account that by the Second Fundamental Theorem of Nevanlinna

$$n_1(cr) \leq \frac{1}{\ln c} N_1(c^2 r) \leq \frac{2}{\ln c} \left[T(c^2 r) + O\left(m\left(c^2 r, \frac{w'}{w} \right) \right) \right]$$

$$\leq \frac{2}{\ln c} T(c^3 r) + o\left(T(c^3 r) \right) \leq \frac{3}{\ln c} T(c^3 r), \quad r > r_1 \tag{4.7.10}$$

and also that $A(cr) \leq (\ln c)^{-1} T(c^3 r)$, we obtain that for $r > r_2$

$$\int_r^{cr} \overline{C}(t, \Gamma)\,dt < K(c)T(c^3 r)(c-1)r + h \int_r^{cr} L(ct)\,dt \leq K(c)rT(c^3 r)$$

(for proving the last inequality we applied the inequality (2.4.1) of Section 2.4 to the integral of the spheric length). Hence, as $\overline{C}(r, \Gamma)$ is an increasing function of r,

$$\overline{C}(r, \Gamma) \leq K(c)T(c^3 r),$$

whence by Lemma 6 of Section 2.4 we obtain the desired inequality (4.7.1).

4.7.6. Proof of Theorem 2. Let the curve $l_\mu^0(r, \Gamma)$ intersecting with $|z| \leq r$ be a pre-image of an arc $\mathfrak{m}_\mu^0(\Gamma) \in \mathfrak{M}_2(\Gamma)$. Keeping the notation used in the previous proof of Theorem 1, one can see that our $l^0_\mu(r,\Gamma)$ is not of the type (ii). Hence

$$C(r,\Gamma) \leq N_1(r,\Gamma) + N_3(r,\Gamma).$$

Obviously by (4.7.10)

$$\sum_{\nu=1}^{q} N_1(r,\Gamma_\nu) \leq 2n_1(cr) \leq \frac{3}{\ln c}T(c^3 r)$$

and by (4.7.14) of Section 4.6

$$\sum_{\nu=1}^{q} N_3(r,\Gamma_\nu) \leq \frac{1}{(c-1)r}\sum_{\nu=1}^{q}[L(cr,\Gamma_\nu) + L(r,\Gamma_\nu)] \leq K(c)T(cr).$$

From the last three inequalities we get

$$\sum_{\nu=1}^{q} C(r,\Gamma_\nu) \leq K(c)T(c^3 r).$$

Applying Lemma 6 of Section 2.4 we obtain the desired inequality (4.7.2).

4.8. The Proximity Property of a-points of Meromorphic Functions

4.8.1. For a long time many successors of Nevanlinna and Ahlfors' theories continued investigations related to the number of a-points of meromorphic functions in $\mathbb{C}$. Now we come to a stage when the questions of locations of a-points, i.e. of their arguments, become of primary interest. Nonetheless, so far there were no general results in this field, and this was compensated by numerous investigations of the arguments of a-points of various subclasses of meromorphic functions.[16]

In the author's article [2] instead of the numbers of the a-points complex sums of a-points were considered which contain natural combinations of modules and arguments of a-points. For these complex sums some results similar to the fundamental theorems of Nevanlinna were established that are true for all meromorphic functions in $\mathbb{C}$. These results show that there is certain closeness between the complex sums of "good" a-points and b-points. Thus,

---

[16]See Bieberbach [1]; Nevanlinna [4]: Edrei [1]: Edrei and Fuchs [1], [2]; Edrei, Fuchs and Hellerstein [1]; Goldberg [2]; Ostrovskii [1, 2].

for all functions meromorphic in $\mathbb{C}$ Nevanlinna's theory provides conclusions on the closeness between the numbers of "good" a-points and b-points, while the above mentioned results provide additional conclusions on the closeness between their complex sums. Consequently, one can presuppose that there is a general phenomenon of closeness that determines a geometrical closeness between "good" a-points and b-points.

We remark that such a phenomenon cannot be derived directly from the proximity of the complex sums of a-points, since it is easy to choose two groups of complex numbers with "close" sums, such that any two representatives of these groups are not geometrically "close" to each other.

The next Theorem 1 reveals a similar phenomenon, which we call Proximity Property of a-points. As a corollary it implies the proximity of the complex sums.

We shall use the following notation. We use $z_i(a)$ for a-points of a meromorphic function $w(z)$, we say $\delta_{a,b} = \delta_{a,b}(r)$ is equal to a if $n(r,a) > n(r,b)$, and is equal to b if $n(r,a) \leq n(r,b)$. Also we denote by $\eta = \eta(r,a,b)$ an enumeration of a- and b-points in the disk $|z| \leq r$ (where a point of a multiplicity k is counted k times). Also we set

$$B_\eta(r,a,b) = \sum |z_i(b) - z_i(a)|, \qquad (4.8.1)$$

where the pairs $z_i(a)$ and $z_i(b)$ ($|z_i(b)| \leq r$, $|z_i(a)| \leq r$) are taken in accordance with the enumeration $\eta(r,a,b)$ and i changes from 1 to $\min\{n(r,a), n(r,b)\}$ (we do not define $B_\eta(r,a,b)$ when $n(r,a) = 0$ or $n(r,b) = 0$). Further, we denote

$$A_\eta(r,a,b) = \sum |z_i(\delta_{a,b})|, \qquad (4.8.2)$$

where $z_i(\delta_{a,b})$ ($|z_i(\delta_{a,b})| \leq r$) are the roots which are not taken into account in (4.8.1) and i is changed from $\min\{n(r,a), n(r,b)\} + 1$ to $\max\{n(r,a), n(r,b)\}$ (we do not define $A_\eta(r,a,b)$ for $n(r,a) = n(r,b)$).

The geometric meaning of the magnitudes A_η and B_η is simple: $B_\eta(r,a,b)$ is the sum of distances between $z_i(a)$ and $z_i(b)$ taken in accordance with the enumeration η, and this sum is a natural measure of the proximity of these pairs. On the other hand, $A_\eta(r,a,b)$ is not zero when the numbers of a- and b-points in the disk are different. Let, for example, $n(r,a) > n(r,b)$. Then evidently $A_\eta(r,a,b)$ is the sum of modules of the points $z_i(a)$, $i = n(r,b) + 1, \ldots, n(r,a)$, taken for the enumeration η.

Now we turn to the main result of this section.

Theorem 1. *Let $w(z) \in M$ be any function and let a_ν and b_ν, $\nu = 1, 2, \ldots, q$, be finite collections of different complex values from $\overline{\mathbb{C}}$.*
Then there exist some absolute constants $K < \infty$ and $C \in (0, 1)$, and for any r there exists a numeration $\eta(r, a_\nu . b_\nu)$ such that

$$\sum_{\nu=1}^{q} [A_\eta(r, a_\nu, b_\nu) + B_\eta(r. a_\nu . b_\nu)] \leq K r A(r) \qquad (4.8.3)$$

for $r \in E$, where $E \subset [0, \infty)$ is some set of lower logarithmic density C.

This theorem gives a general estimate of the distances between the points $z_i(a_\nu)$ and $z_i(b_\nu)$, $\nu = 1, 2, \ldots, q$. forming pairs at the enumeration η. Also the theorem gives an estimate of the sum of modules of $z_i(\delta_{a_\nu, b_\nu})$ that are used in the definition of $A_\eta(r, a_\nu, b_\nu)$. There are some simple examples where (4.8.3) becomes an equality for some values of K.

The below arguments show that Theorem 1 implies the above mentioned proximity phenomenon of a-points. We shall always assume $\varepsilon = \text{const} < 1$ to be small enough. In Theorem 1 we choose the number q of pairs a_ν, b_ν to be equal to $[\varepsilon^{-2}k + 1]'$. Then, except for no more than $[2k\varepsilon^{-1}]'$ pairs, i.e. *for the majority of pairs a_ν, b_ν* we have

$$A_\eta(r, a_\nu, b_\nu) + B_\eta(r. a_\nu . b_\nu) \leq \varepsilon r A(r). \qquad (4.8.4)$$

At the same time, by the Second Fundamental Theorem of Ahlfors for such pairs a_ν, b_ν, the average numbers of a_ν- and b_ν-points are not less than $(1 - \varepsilon)A(r)$. Consequently, from (4.8.4) we conclude that for the "majority" of values a_ν and b_ν and for the "majority" of a_ν-points $z_i(a_\nu)$ and b_ν-points $z_i(b_\nu)$

$$|z_i(a_\nu) - z_i(b_\nu)| \leq \sqrt{\varepsilon} r. \qquad (4.8.5)$$

The last estimate shows that the modules of the a_ν- and b_ν-points are close to each other as well as their arguments. i.e. the a_ν- and b_ν-points are geometrically close to each other. Thus we come to the above mentioned "Proximity Property of a-points" of meromorphic functions.

We remark that the estimate (4.8.5) is not the best for the class of all meromorphic functions. The author established other versions of the Proximity Property [12–13] with the estimate

$$|z_i(a_\nu) - z_i(b_\nu)| \leq \text{const} \frac{r}{A^{1/2}(r)},$$

which cannot be sharpened further and actually contains a series of well-known results including those of Nevanlinna and Ahlfors. Nonetheless, Theorem 1 in essence does not lose its significance because of the simplicity and geometric clarity.

To clarify the enumeration $\eta(r, a_\nu, b_\nu)$ used in Theorem 1 we describe a way to perform such an enumeration. Let Γ_ν be a bounded, smooth Jordan curve in the w-plane, such that: $\nu(\Gamma_\nu) < \infty$, its endpoints are a_ν and b_ν, the set $w^{-1}(\Gamma_\nu) \cap \{z : |z| \leq r\}$ contains exactly $k(r)$ curves γ_i, $i = 1, 2, \ldots, k(r)$, completely contained in $\{z : |z| \leq r\}$ and consisting only of ordinary points of the function $w(z)$. Note that the w-images of such curves are simple islands over Γ_ν (see Section 2.1.10). It is clear that the endpoints of such a curve are a point a_ν and a point b_ν. We assign to these a_ν and b_ν the same index i, i.e. we write them as $z_i(a_\nu)$ and $z_i(b_\nu)$ respectively. So we numerated $k(r)$ pairs $z_i(a_\nu)$, $z_i(b_\nu)$, $i = 0, 1, \ldots, k(r) \leq \min\{n(r, a_\nu), n(r, b_\nu)\}$. The enumeration of the remaining points a_ν and b_ν is arbitrary.

Now observe that for any numeration η

$$|B(r, a) - B(r, b)| \leq A_\eta(r, a, b) + B_\eta(r, a, b). \tag{4.8.6}$$

This inequality and Theorem 1 imply the analog of the Second Fundamental Theorem for the complex sums of a-points, stated in the above mentioned author's paper [2].

4.8.2. Proof of Theorem 1. Let $n_0(r, \Gamma)$ be the number of simple islands of the surface $\widetilde{F}_r$, which lie over Γ (for definition see Section 2.1), and let $\widetilde{n}(r, \Gamma)$ be the number of multiple islands of $\widetilde{F}_r$ (where any term is counted only once). Then evidently

$$n(r, \Gamma) = n_0(r, \Gamma) + n_1(r, \Gamma) + \widetilde{n}(r, \Gamma),$$

whence by the First Fundamental Theorem of Ahlfors we obtain

$$A(r) - n(r, \Gamma) + n_1(r, \Gamma) = A(r) - n_0(r, \Gamma) - \widetilde{n}(r, \Gamma).$$

Hence it follows that the inequality (2.1.34′) of Section 2.1.10 in Ahlfors' second fundamental theorem can be replaced by

$$\sum_{\nu=1}^{q} A(r) - n_0(r, \Gamma_\nu) - \widetilde{n}(r, \Gamma_\nu) \leq 2A(r) + h(\Gamma_1, \Gamma_2, \ldots, \Gamma_q)L(r), \tag{4.8.7}$$

where $h(\Gamma_1, \Gamma_2, \ldots, \Gamma_q) = \text{const} < \infty$. On the other hand, using the obvious inequality $\tilde{n}(r, \Gamma) \leq n_1(r, \Gamma)$ and Ahlfors' First and Second Fundamental Theorems we get

$$\sum_{\nu=1}^{q} [\tilde{n}(r, \Gamma_\nu)] \leq 2A(r) + h(\Gamma_1, \Gamma_2, \ldots, \Gamma_q)L(r).$$

Hence by (4.8.7)

$$\sum_{\nu=1}^{q} [A(r) - n_0(r, \Gamma_\nu)] \leq 4A(r) + h(\Gamma_1, \Gamma_2, \ldots, \Gamma_q)L(r). \qquad (4.8.8)$$

Let Γ_ν, $\nu = 1, 2, \ldots, q$, be the curves defined in Theorem 2 of Section 4.2 and suppose that these curves are chosen such that their endpoints are a_ν and b_ν, $\nu = 1, 2, \ldots, q$. Then denoting

$$n^*(r, a_\nu) := n(r, a_\nu) - n_0(r, \Gamma_\nu)$$

from the inequality (4.8.8) we get

$$\sum_{\nu=1}^{q} n^*(r, a_\nu) \leq 4A(r) - \sum_{\nu=1}^{q} [A(r) - n(r, a_\nu)] + h(\Gamma_1, \Gamma_2, \ldots, \Gamma_q)L(r).$$

Miles showed in [1] (see also Barsegian [5]) that

$$\sum_{\nu=1}^{q} |A(r) - n(r, a_\nu)| \leq KA(r), \qquad r \to \infty, r \in E^*,$$

where K and E^* are the same as in Remark in Section 2.4.1. Hence follows

$$\sum_{\nu=1}^{q} [n^*(r, a_\nu) - n^*(r, b_\nu)] \leq (8 + 2K)A(r)$$

$$+ h(\Gamma_1, \Gamma_2, \ldots, \Gamma_q)L(r), \qquad r \to \infty, r \in E^*.$$
$$(4.8.9)$$

Let now over Γ_ν lie $K_\nu(r)$ simple islands of the surface $\tilde{F}_r$, and let $\gamma_i(\nu)$, $i = 1, 2, \ldots, K_\nu(r)$, be their pre-images. By definition we have $K_\nu(r) \leq \min\{n(r, a_\nu), n(r, b_\nu)\}$ and $\gamma_i(\nu)$ is completely contained in $|z| \leq r$. Further, the endpoints of any $\gamma_i(\nu)$ are an ordinary (not multiple) point a_ν and an ordinary point b_ν, to both we assign the index i, i.e. we write them as $z_i(a_\nu)$ and $z_i(b_\nu)$ respectively. In this way we numerate $K_\nu(r)$ pairs of a_ν- and b_ν-points. Since $|z_i(a_\nu) - z_i(b_\nu)|$ is less than the length of the curve $\gamma_i(\nu)$, and the sum of lengths of all $\gamma_i(\nu)$ is less than $L(r, \Gamma_\nu)$ we come to

$$\sum_{i=1}^{K_\nu(r)} \sum_{\nu=1}^{q} |z_i(a_\nu) - z_i(b_\nu)| \leq L(r, \Gamma_\nu), \qquad (4.8.10)$$

Arbitrarily numerating the remaining a_ν- and b_ν-points for all ν we come to a numeration $\eta(r, a_\nu, b_\nu)$, $\nu = 1, 2, \ldots, q$.

It is obvious that

$$\sum_{i=K_\nu(r)+1}^{\min\{n(r,a_\nu),n(r,b_\nu)\}} |z_i(a_\nu) - z_i(b_\nu)| + A_\eta(r, a_\nu, b_\nu) \leq r\left[n^*(r, a_\nu) + n^*(r, b_\nu)\right],$$

$$(4.8.11)$$

and also it is obvious that

$$A_\eta(r, a_\nu, b_\nu) + B_\eta(r, a_\nu, b_\nu) = \sum_{i=1}^{K_\nu(r)} |z_i(a_\nu) - z_i(b_\nu)|$$

$$+ \sum_{i=K_\nu(r)+1}^{\min\{n(r,a_\nu),n(r,b_\nu)\}} |z_i(a_\nu) - z_i(b_\nu)| + A_\eta(r, a_\nu, b_\nu).$$

Hence by the inequalities (4.8.9)–(4.8.11) we have for $r \in E^*$

$$\sum_{\nu=1}^{q} [A_\eta(r, a_\nu, b_\nu) + B_\eta(r, a_\nu, b_\nu)] \leq \sum_{\nu=1}^{q} L(r, \Gamma_\nu)$$

$$+ (8 + 2K)\Gamma A(\Gamma) + h(\Gamma_1, \Gamma_2, \ldots, \Gamma_q) r L(r).$$

$$(4.8.12)$$

Now taking into account the inequality (11') of Section 4.1.5 we obtain

$$\sum_{\nu=1}^{q} [A_\eta(r, a_\nu, b_\nu) + B_\eta(r, a_\nu, b_\nu)] \leq 2K \int \int_{|z|} \left| \frac{w''(z)}{w'(z)} \right| d\sigma$$

$$+ (8 + 2K) r A(r)$$

$$+ O\left(rL(r) + \int_0^r L(t)dt + 1 \right),$$

which is true for the same $r \in E^*$. Applying Lemma 1' of Section 2.4 and taking into account the Remark to this Lemma 1' we get

$$\sum_{\nu=1}^{q} [A_\eta(r, a_\nu, b_\nu) + B_\eta(r, a_\nu, b_\nu)]$$

$$\leq (8 + 4K)A(r) + O\left(rL(r) + \int_0^r L(t)dt + 1\right),$$

which is true for the same $r \in E^*$ and taking also into account that by Ahlfors' Lemma in Section 2.1 we have $L(r) = o(A(r))$ for $r \to \infty$, $r \in E_0$, where E_0 has finite logarithmic measure and $\int_0^r L(t)dt = o(A(r))$, $r \to \infty$ we obtain Theorem 1 by denoting $K = 8 + 4K + 1$ and $E = E^* \setminus E_0$; the last set, obviously, also has lower logarithmic density C.

4.9. A Proof of the Proximity Property of a-points Based on Investigation of Γ-lines only

3.9.1. Here we give another realization of the program proposed in Section 1.5.3. To do so we need some estimates for the number of connected components of the Γ-lines along with the estimates of Γ-lines. It turns out that the simultaneous use of both these estimates leads to a version of the Proximity Property.

Namely, the following theorem is true.

Theorem 1 *Let* $w(z) \in M$ *be any function with* $\lambda < \infty$, *let* Γ *be a smooth Jordan curve with* $\nu(\Gamma) < \infty$, *and let* $n_r(\Gamma)$ *be the number of connected components of the set of* Γ*-lines in the disk* $D(r)$. *from which all the multiple points are excluded.*

Then there exist constants $K_1(\lambda) < \infty$ *and* $K_2(\lambda) < \infty$ *such that*

$$n_r(\Gamma) \leq K_1(\lambda)T(r) \tag{4.9.1}$$

and

$$L(r, \Gamma) \leq K_2(\lambda)rT(r) \tag{4.9.2}$$

on a sequence $r = r_n \to \infty$.

This theorem guarantees the existence of a sequence $r_n \to \infty$ and some enumerations $\eta = \eta(r_n, a_\nu, b_\nu)$ of pairwise different complex numbers $a_1, a_2, \ldots, a_q$ and $b_1, b_2, \ldots, b_q$ such that

$$\sum_{\nu=1}^{q} \{A_\eta(r_n, a_\nu, b_\nu) + B_\eta(r_n, a_\nu, b_\nu)\} \leq K r_n T(r_n), \qquad (4.9.3)$$

where $K = K(\lambda) < \infty$ is a constant depending solely on λ.

To show this we suppose that: $\gamma_i(r)$, $i = 1, 2, \ldots, n_r(\Gamma)$, are the connected components discussed after the statement of Theorem 1 in the previous Section 4.8; Γ is a curve from the same theorem, on which the mentioned a_ν and b_ν lie in the order $a_1, b_1, a_2, b_2, \ldots, a_q, b_q$; $n_{a_\nu}^*(r)$ (or $n_{b_\nu}^*(r)$) are the numbers of all a_ν-points (or b_ν-points) lying on those $\gamma_i(r)$ which does not contain b_ν-points (or a_ν-points). Taking into account the disposition of a_ν and b_ν on Γ we conclude that

$$\sum_{\nu=1}^{q} n_{a_\nu}^*(r) \leq n_r(\Gamma)$$

and

$$\sum_{\nu=1}^{q} n_{a_\nu}^*(r) \leq n_r(\Gamma).$$

Consequently, from (4.9.1) we find

$$\sum_{\nu=1}^{q} \{n_{a_\nu}^*(r) + n_{b_\nu}^*(r)\} \leq 2 K_1(\lambda) T(r) \quad r = r_n \to \infty. \qquad (4.9.4)$$

Therefore the remaining a_ν-points are such that along with any $a_\nu \in \gamma_j(r)$ the curve $\gamma_j(r)$ contains also a b_ν-point. We assign to all these a_ν- and b_ν-points the same indices (i.e. we denote them $z_{c,j}(a_{\nu_c})$ and $z_{c,j}(b_{\nu_c})$) and then we choose indices of the remaining a_ν- and b_ν-points arbitrarily. As the result we obtain a enumeration $\eta = \eta(r, a_\nu, b_\nu)$. Suppose there are $k(j)$ $(0 \leq k(j) \leq q)$ pairs of a_ν- and b_ν-points $(c = 1, 2, \ldots, k(j))$ on $\gamma_i(r)$. Then by (4.9.2)

$$\sum_{j} \sum_{c=1}^{k(j)} |z_{c,j}(a_{\nu_c}) - z_{c,j}(b_{\nu_c})| \leq K_2(\lambda) r_n T(r_n) \qquad (4.9.5)$$

for $r = r_n$. Now (4.9.4) and (4.9.5) imply (4.9.3) since

$$\sum_{\nu=1}^{q} [A_\eta(r_n, a_\nu, b_\nu) + B_\eta(r_n, a_\nu, b_\nu)]$$

$$\leq \sum_j \sum_{c=1}^{k(j)} |z_{c.j}(a_{\nu_c}) - z_{c.j}(b_{\nu_c})| + r_n \sum_{\nu=1}^{q} \left[n^*_{a_\nu}(r_n) + n^*_{b_\nu}(r_n) \right].$$

4.9.2. Proof of Theorem 1. We split the curves $\gamma_i(r)$ of our theorem into the following three classes.

(i) The curves $\gamma_i(r)$ having no intersections with the circle $|z| = r$ and at least one multiple point as an endpoint. Their number we denote by $N_1^*(r, \Gamma)$.
(ii) The curves $\gamma_i(r)$ having no intersections with the circle $|z| = r$ and no multiple points as endpoints. Their number we denote by $N_2^*(r, \Gamma)$.
(iii) The curves $\gamma_i(r)$ intersecting with the circle $|z| = r$. Their number we denote by $N_3^*(r, \Gamma)$.

It is obvious that

$$N_1^*(r, \Gamma) \leq 2n_1(r) \tag{4.9.6}$$

and

$$N_3^*(r, \Gamma) \leq \Phi(r, \Gamma), \tag{4.9.7}$$

where $\Phi(r, \Gamma)$ is the number of those points z_i on $|z| = r$, the images of which lie on Γ, and $n_1(r)$ is the sum of the orders of the multiple points in $D(r)$.

Any curve $\gamma_i(r)$ of (ii) is a pre-image of a simple island (of multiplicity 1) of the surface $\widetilde{F}_r$, lying over Γ. Consequently, by the First Fundamental Theorem of Ahlfors

$$N_2^*(r, \Gamma) \leq A(r) + h(\Gamma)L(r). \tag{4.9.8}$$

where $h = \text{const} < \infty$. Further, it is obvious that

$$n_r(\Gamma) = N_1^*(r, \Gamma) + N_2^*(r, \Gamma) + N_3^*(r, \Gamma),$$

whence by (4.9.5)–(4.9.8)

$$\int_{r/2}^{r} n_t(\Gamma)dt \leq rn_1^0(r) + \frac{r}{2}A(r) + h\int_{r/2}^{r} L(t)dt + \int_{r/2}^{r} \Phi(t, \Gamma)dt.$$

Taking into account the inequality (4.6.14) of Section 4.6.2 by a standard argument we derive

$$\int_{r/2}^{r} n_t(\Gamma)dt \leq KrT(2r), \quad r > r_1. \tag{4.9.9}$$

On the other hand, by the inequality (4.1.11′) of Section 4.1.5 and Lemmas 1 and 5 of Section 2.4

$$L(r,\Gamma) \leq KrT(2r), \quad r > r_2, \tag{4.9.10}$$

where K is an absolute constant. In accordance with Lemma 6 of Section 2.4 $T(2r) \leq K(\lambda)T(r/2)$ on a sequence $r = r'_n \to \infty$. Therefore, (4.9.9) implies that for any $r'_n > r_2$ there exists a point $r_n \in [r'_n/2, r'_n]$ for which

$$n_{r_n}(\Gamma) = \frac{2}{r'_n} \int_{r'_n/2}^{r'_n} n_t(\Gamma)dt \leq kT(2r'_n) \leq K(\lambda)T(r'_n/2) \leq K_1(\lambda)T(r_n).$$

Since $L(r,\Gamma)$ is an increasing function, from (4.9.10) it follows that for these r_n

$$L(r_n,\Gamma) \leq L(r'_n,\Gamma) \leq Kr_nT(2r'_n) \leq KK(\lambda)r_nT(r'_n/2) \leq K_2(\lambda)T(r_n).$$

This completes the proof of Theorem 1.

CHAPTER 5

SOME APPLIED PROBLEMS

5.1. Γ-Lines in Physics

5.1.1. This monograph is not aimed to discuss in detail the applications of methods of Γ-lines in concrete physical problems but nevertheless briefly outlines some possible fields of such applications, put some new problems turned to applications and give some related methods. In a way, physical phenomena which can be studied by methods of Γ-lines are comparable with those that can be reduced to topological investigation of curves determined by means of differential equations. Topological description of such curves is connected with the property of locality of a physical process in a bounded part of the "space". The latter is one of the fundamental ideas of Poincare. An example of physical treatment of such a topological fact is locality of planets, i.e., usability of planets is contained to come out in a bounded part of the Universe.

Following this idea we interpret Γ-lines as the boundaries of domains in which an extremal for a given process or physical phenomenon appears or as the lines indicating the locality of a physical process. We apply a metrical approach instead of a topological one. It is known that many physical processes are described by harmonic or more general functions $\nu(x, y)$ indicating, for example, the temperature in a point (x, y) of a thermal phenomena, the velocity of the movement of a point in hydrodynamics, the tension in the theory of elasticity and electric processes etc. For a given function $\nu(x, y)$ we introduce a complex function $w(z) = u(x, y) + i\nu(x, y)$. If $\gamma(0)$ is the real axis, then the $\gamma(0)$-line of the function $w(z)$ is a level set of $\nu(x, y)$, in which the temperature, the movement velocity or the tension is equal to zero. Similarly, if $\gamma(t)$ is the straight line $\{w : \operatorname{Im} w = 0\}$, then $\nu(x, y) = t$ on a $\gamma(t)$-line. Hence a $\gamma(t)$-line separates the domains in which the mentioned physical magnitudes are correspondingly greater or less than t. Consequently, while investigating $\gamma(t)$-lines for the "critical" values of t, we, actually, investigate the boundaries of the domains where the considered physical process becomes extremal or critical. In other words, a physical phenomenon undergoes a "catastrophe" on $\gamma(t)$-lines. It is clear that: the domains of high tension are the most possible places for the cracking of the material in the theory of elasticity, the domains of high voltage are the most possible places for electric discharges in electrodynamics and the domains of high temperature are where substance turns to plasma. $\gamma(t)$-lines for $t = 0°C$ and $t = 100°C$ can be considered as lines separating the domains of ice, water and steam for thermal processes in water. It is evident that for a critical value t the smallness of the length of critical $\gamma(t)$-lines in comparison with the whole boundary of the given process implies that the extremal behavior is observed only in some small parts of the field, i.e. the phenomenon is localized in a certain sense. Also, the smallness of the boundary of domains of high tension and temperature can lead to the conclusion that, for example, the cracks inside the material are localized and cannot spread down to the boundary and that steam, magma and plasma conceived in the interior of the earth will not spread outside.

Thus, a wide field of physical problems can be distinguished for which the evaluation of the lengths of Γ-lines is clearly useful. Hence the tangent variation principle can be applied for the investigation of these problems.

It is an interesting fact that if $w \in M_R$ and the function $\nu(x,y) = \operatorname{Im} w$ is describing a given physical process, then estimates of the length of $L(r, \Gamma)$, $r < R$, $\Gamma = \{w : \operatorname{Im} w = A = \text{const}\}$ (i.e. the lengths of the curves on which $\nu(x,y) = A$) can be rewritten completely in terms of the function $\nu(x,y)$. Namely, in the form

$$
\begin{aligned}
L(r,\Gamma) &\leq \frac{4\sqrt{2}}{\pi} \int\int_{|z|<r} \frac{|w''(z)|}{|w'(z)|}\, d\sigma + 2\sqrt{2}\pi r \\
&\leq \frac{4\sqrt{2}}{\pi} r' \ln\frac{\rho^2}{\rho^2 - r^2} \int_0^{2\pi} \left|\ln\left|\operatorname{grad}\nu(\rho^{i\varphi})\right|\right| d\varphi + 32\sqrt{2}r(n_0 + n_\infty + 1),
\end{aligned}
$$
(5.1.1)

where $r < \rho < R$, n_0 and n_∞ are the numbers of the critical points (x,y) lying in $|z| < \rho$, where $\operatorname{grad}\nu(x,y) = 0$ and $\operatorname{grad}\nu(x,y) = \infty$ respectively. This immediately follows from the tangent variation principle, proof of Lemma 1 of Section 2.4 and from $|w'| = |\operatorname{grad}v(x,y)|$.

We now put forward a new problem:[17] to describe solutions of the equation

$$
F(\nu, \nu_x', \nu_y', \dots, \nu_{xx}'', \nu_{yy}'', \dots, x, y) = 0
$$
(5.1.2)

---

[17]Other problems related to Γ-lines can be seen in a recent collection of open problems by Beardon and Yang [1].

in terms of the function $\nu(x, y)$. In the case when $F = \nu - K$, K is a constant, these solutions are usual level sets. Obviously, these solutions can have a much wider area of applications.

Denote by $L_F(r)$ the length of the curves which are the solutions of $F = 0$. Particularly when F is the real (or imaginary) part of a new function W meromorphic in $\mathbb{C}$ or in the unit disk then due to (5.1.1) we have

$$L_F(r) \leq \frac{4\sqrt{2}}{\pi} \int \int_{|z| < r} \frac{|W''(z)|}{|W'(z)|} d\sigma + 2\sqrt{2}\pi r, \qquad (5.1.1')$$

so that if the characteristic $T(r, W)$ can be estimated by the characteristic $T(r, w)$ then we can get bounds of $L_F(r)$ in terms of $T(r, w)$. In the interesting case when F is a linear combination of derivatives of $\nu = \operatorname{Re} w$, $w \in M$, with respect to x and y, we have a constant K_1 depending only on this combination

$$T(r, F) < K_1 T(r, w), \quad r > r_0,$$

so that making use of (5.1.1') and Lemma 1 of Section 2.4 we easily obtain a constant $K(c)$, depending only on $c = \text{const} > 1$:

$$L_F(r) \leq K_1 K(c) T(cr, w), \quad r > r_0. \qquad (5.1.3)$$

Remark. In the case when w is a univalent function in the unit disk, the problem generalizes the known problem on bounds of $L(r, \Gamma)$ considered by numerous authors (see Subsection 1.2.1). The approach used above can be used to get bounds of $L_F(r)$ in this case as well. However here we cannot directly make use of the results of these authors since linear combinations of derivatives of $w(z)$ are not more univalent functions in general. But we again can make use of the Tangent Variation Principle since it is true for arbitrary meromorphic functions.

5.2. On the Cross Road of Value Distribution, Γ-lines, Free Boundary Theories and Applied Mathematics

5.2.1. Introduction. Our aim[18] is to establish a new mathematical topic which admits numerous interpretations in physics, physical chemistry, and ecology and which applies ideas, concepts, methods and results of classical

---

[18]This section is due to a joint work by professor Yang and the author, see Barsegian and Yang [1].

Nevanlinna Value Distribution Theory and Γ-lines' Theory for the quantitative descriptions of solutions of the following system:

$$v(x, y) = \text{const} = t, \ |\text{grad}\, v(x, y)| = \text{const} = R. \tag{5.2.1}$$

The existence of the solutions has been considered by many experts in the past two decades in the studies of Free Boundary Theory. However, we now consider quantity of these solutions and transfer the classical problems of Value Distribution Theory for a-points and those considered above for Γ-lines into the field of solutions of (5.2.1). For these quantities we establish a regularity that surprisingly resembles the Second Fundamental Theorem of Nevanlinna and his Deficiency Relations.

The classical concepts and result transferred into new situation acquire important physical interpretations due to the rich applied content of Free Boundary Theory and Γ-lines Theory.

Similar to equation (5.2.1), we have:

$$v(x, y) = \text{const} = t, \tag{5.2.2}$$

whose solutions are particular cases of Γ-lines, and also similar to equation (5.2.1), we have

$$v(x, y) = t = \text{const}, \ u(x, y) = R = \text{const}. \tag{5.2.3}$$

whose solutions are a-points of the complex function $w(z) = u(x, y) + iv(x, y)$ when $a = t + iR$.

Despite the outward similarity between equation (5.2.1) and equation (5.2.3), the corresponding solutions may have strikingly different behaviors. However, our feeling was that there must be also some regularities connecting the solutions to (5.2.1)–(5.2.3).

Indeed, it appears that the Γ-lines' Theory is a tool which enables us to describe the numbers of solutions of (5.1.1) and leads to the above mentioned analog of the Nevanlinna Theory for solutions of (5.1.1).

5.2.2. On solutions of system (1): physical interpretations. Solutions of equation (5.2.1) for different classes of functions $v(x, y)$ of two variables are of great interest in applied mathematics. They are also "hot topics" in physics, particularly in meteorology and environmental problems. These solutions are interpreted, for instance, as sets where the temperature, level of radiation, velocity, level of oil on a water surface and so on are equal to a given constant t and, simultaneously, the growth of these temperatures and so on in these sets (which is described by $|\mathrm{grad}\,v(x,y)|$) is equal to a given constant R.

Actually solutions described can be interpreted as "propagations of level sets in physical processes" or propagations that have been for instances of fare, flame, radiation, oil soil and so on.

These solutions studied for different classes of $v(x,y)$ for a long time constitute at present the so-called Free Boundary Problems (or Free Boundary Theory). Thanks to the initial work by Beurling [1] in the 1950s and the applied aspects showed by Lions [1], Zeldovich and his co-workers Frank-Kamenetskii, Barenblatt, Libovich, Makhviladze [1–2] and more recent investigations by Caffarelli and co-workers (see for instance Alt and Caffarelli [1] and Caffarelli and Vazquez [1] and references there), this theory has numerous followers in many countries at present. In the above mentioned pioneering works in the Free Boundary Problems one can find more than 60 articles in physics and physical chemistry where these solutions play a crucial role.

Free Boundary Problem requires a harmonic function in the given domain D $v(x,y)$ and a free domain $d \subset D$ such that $v(x,y) = 1$ on the boundary of D and simultaneously with $v(x,y) = 0$, and $|\mathrm{grad}\,v(x,y)| = 1$ on the boundary of d. In diverse modifications and generalizations of this problem, harmonic functions were replaced by other functions (for instance, solutions of heat equations) and were considered as functions in the space.

5.2.2. Subject of the new proposed topic: movement of problems and ideas.
Unlike a Free Boundary Theory which considers the existence of the solutions of (5.2.1) for some classes of $v(x,y)$ we propose to give quantitative descriptions of these solutions and to transfer the main concepts and conclusions of the Value Distribution Theory and Γ-lines' Theory into the field of solutions of (5.2.1) for some possibly large classes of $v(x,y)$. Clearly, the classical concepts and results transferred into new situation acquire important physical interpretations due to the rich applied content of the Free Boundary Theory and Γ-lines' Theory.

Our approach is based on the following observation: the solutions of (5.2.1) are actually solutions of complex equations $F(z) = a$, where $F(z) := v(x,y) + i|\mathrm{grad}\,v(x,y)|$ and $a = t + iR$, that is, they are the usual a-points $z := F^{-1}(a)$ of the complex function $F(z)$. Thus the study of solutions of (5.2.1) is reduced to the study of non-analytic functions $F(z)$, and we come to the clear idea: to describe the so-called Value Distribution for the solutions of (5.2.1) as the classical Nevanlinna Value Distribution Theory and Ahlfors' Theory of Covering Surfaces describe the numbers $n(r, a, w)$ of the a-points of meromorphic functions.

Our preliminary results establish connections between the main concepts and magnitudes in these theories and lead to an analog of the classical Deficiency Relation for solutions of (5.2.1).

This shows that we are now at the threshold of a new Value Distribution Theory for solutions of (5.2.1) and the new theory enriches the mentioned theories by mutual penetrations of problems and ideas and by physical interpretations of mathematical concepts.

Note that for different classes of smooth functions $v(x, y)$ the a-points of $F(z)$ can be points as well as lines or even domains.[19]

In a similar way, define the number $n(D, v = t, |\text{grad } v| = R)$ of the a-points $(a = t + iR)$ of the function $F(z)$ in the given domain D or the number of the solutions of (5.2.2) in the domain D. Here, any pointwise solution of (1) in the domain D is counted only once in $n(D, v = t, |\text{grad } v| = R)$ and if a solution of (5.2.1) (which is the same "a-points" of the functions $F(z)$) is a line or a domain belonging to D, we count the line or domain only once in $n(D, v = t, |\text{grad } v| = R)$.

5.2.3. Γ-lines of functions $w(z)$ as a tool for studying the Value Distribution of functions $F(z)$.

Now we make some connections between the length of Γ-lines and the numbers of the solutions of equation (5.2.1) for the important case when $v(x, y)$ is the imaginary part of a function $w(z)$ meromorphic in $\overline{D}$. Then $|\text{grad } v(x, y)| = |w'(z)|$, and hence $F(z) = v(x, y) + i|w'(x + iy)|$. Therefore, the number $n(D, v = t, |\text{grad } v(x, y)| = R)$ of solutions of (5.2.1) in D becomes the number $n(D, v = t, |w'| = R)$ of solutions of the system

$$v(x, y) = t = \text{const}, \quad |w'(x + iy)| = R = \text{const},$$

which, in turn, coincides with the number of $t + iR$-points of $F(z)$ which lies in the domain D.

One can easily prove the following theorem which establishes a connection between the number of the solutions of (5.2.1) and the length of Γ(R)-lines

---

[19]Note that for different classes of smooth functions $v(x, y)$ the a-points of $F(z)$ can be points as well as lines or even domains. Note also that for analytic function $w(z)$ in a domain D the a-points belonging to D indeed are points.

of the derivative w'. We consider averaged magnitudes[20]

$$N(D, R, v) := \int_{-\infty}^{+\infty} n(D, v = t, |\operatorname{grad} v| = R)dt. \qquad (5.2.4)$$

Theorem 1. *Suppose that $v(x, y)$ is the imaginary part of a meromorphic function $w(z)$ in the given domain D. Then*

$$N(D, R, v) \leq RL(D, \Gamma(R), w'). \qquad (5.2.5)$$

The estimates of the number of solutions of (5.2.1) can be given also in terms of integrals S_1, S_2, S, A_1 and A_2 defined in Section 3. Similar estimates can be of interest since all these integrals have a clear physical meaning and appear often in pure mathematical and applied investigations.

Theorem 2. *Under the condition of Theorem 1 the following inequalities*

$$\int_0^{+\infty} \frac{N(D, R, v)}{R} dR \leq S_1(D, w') \leq [S(D)S_2(D, w')]^{1/2} \qquad (5.2.6)$$

and

$$\int_0^{+\infty} \frac{N(D, R, v)}{(1 + R^2)R} dR \leq A_1(D, w') \leq [S(D)A_2(D, w')]^{1/2} \qquad (5.2.7)$$

hold, where S_1, S_2, S, A_1 and A_2 are those defined in Section 3.

Proof. Suppose that $l(|w'| = R)$ are the Γ-lines of the function $w'(z)$ in D. Mapping them by $w(z)$ we obtain a collection of curves $l_w^{w'} := w(l(|w'| = R))$. Observe that by definition we have $|w'| = R$ on the curves $l(|w'| = R)$. Therefore, denoting the length of X by $|X|$ we get

$$|l_w^{w'}| := \int_{l(|w'|=R)} |w'| ds = R|l(|w'| = R)| := RL(D, \Gamma(R), w'). \qquad (5.2.8)$$

Now consider the points $z_i \in D$, where $l(|w'| = R)$ and the $\gamma^*(t)$-lines, $\gamma^*(t) = \{w : \operatorname{Re} w = t\}$ of the function $w(z)$ are intersecting. These actually are the a-points of the function $F(z)$, where $a = t + iR$. The w-images of all points z_i lie on the straight line $\gamma^*(t) = \{w : \operatorname{Re} w = t\}$. Besides, the number of the points $w(z_i)$ (multiplicities counted) is equal to $n(D, v = t, |w'| = R)$.

---

[20]It is similar to the averaged number of a-points considered in the Nevanlinna Theory: $\int_0^r \frac{n(t,a,w)}{t} dt$.

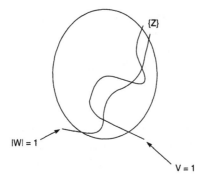

Figure 5.1

Now we integrate this number by t and make use of the geometric reasonings (see Figures 5.1 and 5.2).

Then it is obvious that the integral

$$\int_{-\infty}^{+\infty} n(D, v = t, |w'| = R)dt$$

is equal to the total length of the projections of the set of curves $l_w^{w'}$ to the axis v, $(w = u + iv)$, where the multiplicity is accepted.

In turn, the total length of the projections does not exceed the length $|l_w^{w'}|$ itself. Therefore from (5.2.8) we obtain the inequality (5.2.5) of Theorem 1.

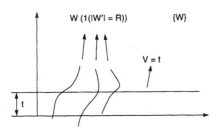

Figure 5.2

The estimate (5.2.5) seems to be exact within a constant. The only place where we were rough was the replacement of the length of the projection of $w(l(|w'| = R))$ by its own length.

Integrating (5.2.5) and applying the estimates (1.1.8) and (1.1.9) of Section 1.1, we come to the desired inequalities (5.2.6) and (5.2.7) of Theorem 2.

5.2.4. An analog for the Second Fundamental Theorem of Nevanlinna for the solutions of system (5.2.1).

Now we suppose that $v(x, y)$ is the imaginary part of a meromorphic function w in the complex plane and R_ν, $\nu = 1, 2, \ldots, q$, are pairwise different real numbers, $0 < R_\nu < \infty$. Then the inequality (5.2.5) implies

$$\sum_{\nu=1}^{q} N(r, R_\nu, v)$$

$$\leq \sum_{\nu=1}^{q} R_\nu L(r, \Gamma(R_\nu), w') \leq \left[\max_\nu R_\nu\right] \sum_{\nu=1}^{q} L(r, \Gamma(R_\nu), w'). (5.2.9)$$

On the other hand due to the inequality (4.1.11') of Section 4.1 we have

$$\sum_{\nu=1}^{q} L(r, \Gamma(R_\nu), w') \leq 2K \int \int_{|z|<r} \left|\frac{w'''(z)}{w''(z)}\right| d\sigma$$

$$+h(\Gamma_1, \Gamma_2, \ldots, \Gamma_q) \int_0^r L(t, w') dt + \sqrt{8\pi r},$$

and hence, arguing similarly as in Section 4.1.5 and using standard reasonings in the Nevanlinna Value Distribution Theory we obtain

$$\sum_{\nu=1}^{q} L(r, \Gamma(R_\nu), w') \leq \frac{K_0}{1 - c} T(cr, w), \quad r = r_n \to \infty, \qquad (5.2.10)$$

where $K_0 < \infty$ is an absolute constant, c is an arbitrary constant > 0.

From (5.2.9) and (5.2.10) we come to the following analog of the Second Fundamental Theorem of Nevanlinna for the solutions of system (5.2.1).

Theorem 3. *Suppose that $v(x, y)$ is the imaginary part of a meromorphic function $w(z)$ in the complex plane and R_ν, $\nu = 1, 2, \ldots, q$, are pairwise* *different real numbers, $0 < R_\nu < \infty$. Then*

$$\sum_{\nu=1}^{q} N(r, R_\nu, v) \leq \left[\max_\nu R_\nu\right] \frac{K_0}{1-c} T(cr, w), \quad r = r_n \to \infty. \quad (5.2.11)$$

Let us consider the connections between Theorem 3 and the Second Fundamental Theorem of Nevanlinna. The main Deficiency Relation in Value Distribution Theory is derived from the circumstance that the right-hand side in the Second Fundamental Theorem, the inequality (2.1.6) of Section 2.1, does not depend on the number of values a_ν considered in this inequality. Due to the classical Nevanlinna Deficiency Relations for meromorphic functions w in the complex plane and for any $a \in \mathbb{C}$, except not more than a countable set of values of a and for exceptional a with $\delta(a) > 0$, we have

$$\sum_{a \in \mathbb{C}} \delta(a) \leq 2.$$

Now, arguing as in the Nevanlinna Value Distribution Theory, we introduce *deficient* values R_ν for the solutions of (5.2.1). We say that a value R_ν is deficient if

$$\tilde\delta(R_\nu) := \lim_{r\to\infty} \inf \frac{N(r, R_\nu, v)}{T(cr, w)} > 0.$$

From (5.2.11) it follows:

Deficiency Relations for solutions of (5.2.1). *Suppose that $v(x, y)$ is the imaginary part of a meromorphic function $w(z)$ in the complex plane. Then for any $R < R_0 = \mathrm{const} > 0$, except not more than a countable set of values R, such that*

$$\tilde\delta(R) = 0 \qquad (5.2.12)$$

and

$$\sum_{0 < R < R_0} \tilde\delta(R) \leq \frac{R_0 K_0}{1-c} < \infty. \qquad (5.2.13)$$

Thus, due to (5.2.12), we come to a surprising conclusion that for the majority of the R values the number of the solutions of (5.2.1) in the disks

$D(r)$ is asymptotically small. This means that deficient R, or which is the
same as those R for which $\tilde{\delta}(R) > 0$, is exceptional. However (5.2.13) gives a
total estimate from above for the number of the solutions of (5.2.1) with these
exceptional or deficient values R. Despite the outward similarity between the
Nevanlinna Deficiency Relations and the Deficiency Relations for solutions
of (5.2.1) we can clearly see also some striking differences between phenomena related to the numbers $N(r, a, w)$ and those related to $N(r, R_\nu, v)$: the
Nevanlinna Deficiency Relations shows that for the majority of $a \in \mathbb{C}$, the
magnitudes $N(r, a, w)$ are asymptotically "big"; the Deficiency Relations for
solutions of (5.2.1) shows for the majority of R, $0 < R < R_0$ the magnitudes
$N(r, v = t, |\text{grad } v| = R_\nu)$ are asymptotically "small".

**5.2.5. Connections between the deficiency relation in the Nevan-
linna theory, the deficiency relation in Γ-lines theory, the deficiency
relation for solutions of (5.2.1).** Note that making use of Theorem 1 we
can give upper bounds for the deficiencies $\tilde{\delta}(R_\nu, w')$ for the solutions of (5.2.1)
in terms of deficiencies $\delta^*(\Gamma(R_\nu), w')$ of the curves $\Gamma(R_\nu)$ of the derivatives
of w', see Section 4.1. On the other hand making use of in Section 4.1 the
established connections between $\delta(a, w)$ and $\delta^*(\Gamma, w)$ (when $a \in \Gamma$) we can
derive bounds from below for $\tilde{\delta}(R_\nu, w')$ in terms of $\delta(a, w')$ (when $a \in \Gamma(R_\nu)$).

We do not have any comprehensive explanation as to why the last conclusion takes place since too little is known about the solutions of (5.2.1) and
too few examples were considered before.

Note that we came to the deficiency relation for solutions of (5.2.1) when
$v(x, y) = \text{Im } w(z)$.

However we expect that there are similar phenomena for some other more
general classes of $v(x, y)$. The goal of the next section is to establish some
methods for the study of similar phenomena in general.

5.2.6. Some open problems. The upper and lower estimates for the numbers of the solutions of (5.2.1) may show how widely the sets of the solutions
can be disseminated. Clearly, in application this means a dissemination of
a given physical phenomenon. Taking also into account how popular the
pure mathematical Value Distribution Theory is and how promising to Free
Boundary Theory is its application, one may expect that the following question can be of great interest: "Is it possible to give quantitative descriptions
for the numbers $n(D, v = t, |\text{grad } v| = R)$ for some large or in applica-
tion important functions $v(x, y)$? Moreover, one may ask: Are there any

regulations similar to the main conclusions of the Value Distribution Theory in the case of solutions to (5.2.1) for some more general classes of functions $v(x, y)$?

The following chain of questions arise now naturally: what can be said in general about the number of the a-points, singularities, about the Value Distribution of the functions $F(z) := v(x, y) + i|\text{grad } v(x, y)|$, or the solutions of system (5.2.1) in the case

1. when $v(x, y)$ is the imaginary part of an analytic function $w(z)$ in the given domain D, particularly if $w(z)$ is entire or meromorphic in the complex plane?

2. when $w(z)$ is a quasiconformal mapping in D or in the complex plane?

3. when $v(x, y)$ is just harmonic in D?

4. when $v(x, y)$ is a solution of the Poisson equation in D?

5. when $v(x, y)$ is a solution of the heat equation in D?

6. when question 1 is considered in $\mathbb{C}^n$ or questions 2–5 are considered in the space $\mathbb{R}^n$?

The Free Boundary Theory considers the existence of solutions of (5.2.1) for specific types of domains D and $v(x, y)$ being solutions of the Laplace or the heat equations and these cases have a well-recognized value for applications due to physical interpretations. We may consider the Value Distribution for corresponding $F(z)$ related to the same D and $v(x, y)$. Then we deal with the same solutions with the same interpretations which, this time, we are going to describe quantitatively. Therefore we are especially interested in the consideration of these cases.

Let us now consider the solutions of systems

$$v(x, y) = t, \quad G(v, v_x', v_y', x, y) = 0$$

with some rather general functions G. Such systems embrace incomparably larger situations in physics than those in the Free Boundary Theory. Therefore we believe that the above problems 1–6 for this system, or the same for the a-points of the functions $F^*(z) = v(x, y) + iG(v, v_x', v_y', x, y)$ can be of interest in many applied topics.

We are especially interested in giving quantitative description for solutions of the following particular case of the above system

$$v(x, y) = t = \text{const}, \quad \arg \text{grad } v(x, y) = \varphi = \text{const},$$

since their solutions, clearly, describe the geometry of the family of curves $v(x, y) = t = $ const. As mentioned above we can consider the solutions as a-points of the following function $\tilde{F}(z) = v(x, y) + i \arg \text{ grad } v(x, y)$. Thus, if we can say something about a-points of $\tilde{F}(z)$, we obtain a much more comprehensive knowledge of the propagations of related physical magnitudes.

By examining even the simplest cases when $v(x, y)$ is an imaginary part of a simple entire function $w(z)$ we easily see that topology and geometry of $w(z)$ and $F(z)$ are strikingly different. All the more the following question is of interest: are there any general results in the classical Value Distribution Theory which, at least with certain changes, are valid also in the case of the functions $F(z)$ or the functions $F^*(z)$? Especially we are interested to learn those properties of the Value Distribution of $F(z)$ which can be described by means of classical concepts in the Value Distribution of functions $w(z)$: since $F(z)$ has certain physical meaning in many topics, this is a way to prescribe a physical meaning to the mentioned classical concepts in the usual Value Distribution Theory (for functions $w(z)$).

Thus we see a certain prospect for constructing a theory directed to divers' applied problems (also those arising in Free Boundary Theory) and resembling the classical Value Distribution Theory.

However, as we have said. there are many differences between behaviors of the functions $w(z)$ and $F(z)$ and to go ahead in constructing this theory we need to first prepare some tools for the study; since none of the many ready tools in the classical Value Distribution Theory works in the case of non-analytic functions $F(z)$.

In the next section we give some methods for the study of the numbers of a-points of general classes of complex functions.

5.3. "Point maps" of Physical Processes and a-points of General Classes of Functions

5.3.1. One can observe that $\gamma(t)$-lines of a function $w(z) = u(x, y) + iv(x, y)$ drawn for some values $t = t_1, t_2, \ldots$ give a "map" of $v(x, y)$ or of the corresponding physical process. If the graph of $v(x, y)$ corresponds to a landscape surface, then we deal with a usual geographic map.

It is possible to offer another "point method" of landscape description. Singling out all those points (x, y) on a plane, where the height $v = v(x, y)$ of the surface is equal to a given value t_1 and the normal n_v to the surface is equal to $R_1, R_2, \ldots$, we obtain a certain description of the contour of a mountain on the height t_1. Besides, varying the heights $t_1, t_2, \ldots$ etc. we characterize the contour of the surface pointwise, i.e. we get a point map of a mountain. Such a description is mathematically reduced to a solution of the system $\nu = t_i$, $n_\nu = R_j$, $i = 1, 2, \ldots, j = 1, 2, \ldots$.

An analog of this idea for physical processes is the description of a physical phenomenon defined by a function $\nu(x, y) = t$ by means of solution of the system

$$F_1(\nu, x, y) = 0, \quad F_2(\nu, x, y) = 0. \tag{5.3.1}$$

Clearly the solutions of the system $\nu(x, y) = t$, $|\operatorname{grad} \nu(x, y)| = R$ considered in Section 5.2 are just very particular case of the system (5.3.1).

Thus, we obtained a usual problem of the Value Distribution Theory, but it is for the function $W(z)$, which is no more analytic. Therefore, we found that the Value Distribution problems for general classes of functions certainly are connected with the description of various physical phenomena. The behavior of zeros and singularities of such general functions and functions with an analytic structure can differ strikingly. As a result, new methods are needed for investigating zeros of the function $W(z) = F_1 + iF_2$ or which is the same as solutions of systems similar to (5.3.1).

5.3.2. Evaluation method for the number of zeros of real functions. Suppose $f(x)$ is a function continuously differentiable twice on a segment $a \leq x \leq b$ and let $x_1, x_2, \ldots, x_n \in [a, b]$ be its zeros.

We introduce a new complex-valued function $F(x) = U(x) + iV(x)$ of the real variable $x \in [a, b]$, satisfying the condition $F(x) = 0$ at the points

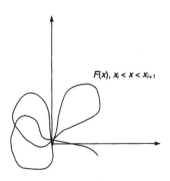

Figure 5.3

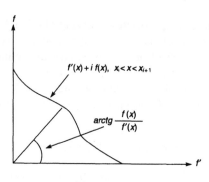

Figure 5.4

$x_1, x_2, \ldots, x_n$. As such this function can take, for example, $F(x) = f(x) + i\varphi(x)f(x)$, where $\varphi(x)$ is such that $\varphi(x) \neq 0$ and $\varphi'(x) = 0$ for $x \in [a, b]$. Obviously, the graph of $F(x)$ is a curve closed at the origin (see Figure 5.3).

Therefore the variation of the angle $\arg \frac{d}{dx} F(x)$ between the tangent to the curve $F(x)$ and the real axis is not less than π on any segment x_i, x_{i+1}, i.e.,

$$\pi \leq \int_{x_i}^{x_{i+1}} \left| \frac{d}{dx} \arg \frac{d}{dx} F(x) \right| dx.$$

Therefore the following estimate is true for the number of zeros of $f(x)$:

$$n \leq \frac{1}{\pi} \int_a^b \left| \frac{d}{dx} \arg \frac{d}{dx} F(x) \right| dx + 1. \tag{5.3.2}$$

If we consider only the number n' of those zeros $x_i' \in [a, b]$ of $f(x)$ for which $f'(x_i') \neq 0$, then (see Figure 5.4)

$$n' \leq \frac{1}{\pi} \int_a^b \left| \frac{d}{dx} \arctan \frac{f(x)}{f'(x)} \right| dx + 1. \tag{5.3.3}$$

To prove this estimate we only have to note that in (x_i', x_{i+1}') there exists a point x^* such that $f'(x^*) = 0$. Hence the angle $\arctan f(x)/f'(x)$ is equal to zero, and the angle $\arctan f(x^*)/f'(x^*)$ is equal to $\pi/2$.

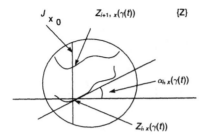

Figure 5.5

Hence

$$\frac{\pi}{2} \le \int_{x_i}^{x^*} \left| \frac{d}{dx} \arctan \frac{f(x)}{f'(x)} \right| dx, \quad \frac{\pi}{2} \le \int_{x^*}^{x'_{i+1}} \left| \frac{d}{dx} \arctan \frac{f(x)}{f'(x)} \right| dx$$

so that

$$\pi \le \int_{x_i}^{x'_{i+1}} \left| \frac{d}{dx} \arctan \frac{f(x)}{f'(x)} \right| dx$$

and summing these estimates over all i we come to (5.3.3).

5.3.3. An evaluation method for integrals $\int_{w^{-1}(\gamma(t))} |\varphi(s)| ds$.

For omitting non-essential details, down to the end of this chapter we assume that all the improper integrals we consider exist. Further, for simplicity we shall assume that the function $w(z)$ is meromorphic in the disk $D(r) = \{z : |z| < r\}$, i.e. $w(z) \in M_{D(r)}$.

Also we shall use the notations $l_x(\Gamma)$, $l_y(\Gamma)$, J_x, J_y, $\Phi(D, x, \Gamma)$, $\Phi(D, y, \Gamma)$, $\alpha_{i,x}(\Gamma)$ and $\alpha_{j,y}(\Gamma)$ from Section 1.1, where $\Gamma = \gamma(t) = \{w : \mathrm{Im}\, w = t\}$ and $D = D(r)$. Suppose that for a given x_0 the function $w(z)$ has no singular points on J_{x_0} and that the number $\Phi(D(r), x_0, \gamma(t))$ of points $z_{i,x}(\gamma(t))$ where the curves $l_x(\gamma(t))$ intersect with J_{x_0} is not less than 2.

Then between any two successive points $z_{i,x}(\gamma(t)) \in J_{x_0}$ and $z_{i+1,x}(\gamma(t)) \in J_{x_0}$ with w-images in $\gamma(t)$ (see Figure 5.5) there exists a point $z^* \in J_{x_0}$ such that the tangent to the curve $w(J_{x_0})$ at $w(z^*)$ is parallel to the straight line

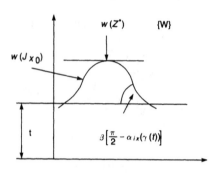

Figure 5.6

$\gamma(t)$ (see Figure 5.6). Therefore for a given real. continuous function $\varphi(z)$ defined in $\overline{D}(r)$ the integral

$$\int \left| \left(\varphi(z) \sin \arg \frac{\partial}{\partial y} w(z) \right)'_y \right| dy \qquad (5.3.4)$$

over J_{x_0}, taken from $z_{i.x_0}(\gamma(t))$ to z^* or from z^* to $z_{i+1,x_0}(\gamma(t))$, is not less than

$$|\varphi(z_{i.x_0}(\gamma(t)))| \sin \beta \left(\frac{\pi}{2} - \alpha_{i.x_0}(\gamma(t)) \right)$$

or

$$|\varphi(z_{i+1,x_0}(\gamma(t)))| \sin \beta \left(\frac{\pi}{2} - \alpha_{i+1,x_0}(\gamma(t)) \right),$$

respectively, where $\beta(\alpha)$ is the w-image of the angle α.

    Besides, as the angles $\alpha_{i.x_0}(\gamma(t))$ and $\alpha_{i+1.x_0}(\gamma(t))$ are not greater than $\pi/4$, using the definition of the class $M_D(\alpha)$ we come to the estimates

$$\sin \beta \left(\frac{\pi}{2} - \alpha_{i.x_0}(\gamma(t)) \right) \geq \sin \alpha,$$

and

$$\sin \beta \left(\frac{\pi}{2} - \alpha_{i+1.x_0}(\gamma(t)) \right) \geq \sin \alpha.$$

Consequently, from (5.3.4) it follows that

$$\sum |\varphi\left(z_{i,x_0}(\gamma(t))\right)|$$

$$\leq \frac{1}{\sin\alpha} \int_{J_{x_0}} \left| \left(\varphi(z) \sin\arg\frac{\partial}{\partial y}w(z) \right)'_y \right| dy + \max_{J_{x_0}} |\varphi(z)|,$$

where the last summand exists only if $\Phi(D(r), x_0, \gamma(t)) = 1$. As $\alpha_{i,x} \leq \pi/4$, at any point $z \in l_x(\gamma(t))$ we have $\Delta s\cos a_{i,x}(\gamma(t)) \geq \frac{\sqrt{2}}{2}\Delta x$ for the length element Δs of the curve $l_x(\gamma(t))$. Therefore,

$$\int_{l_x(\gamma(t))} |\varphi(z)|ds \leq \frac{2}{\sqrt{2}\sin\alpha} \int_{-r}^{r} \int_{J_{x_0}} \left| \left(\varphi(z) \sin\arg\frac{\partial}{\partial y}w(z) \right)'_y \right| dy\,dx$$

$$+ \frac{\sqrt{2}}{2} \int_{-r}^{r} \max_{J_x} |\varphi(z)|dx := \Phi_x\left(r, \varphi(z)\right).$$

In the two previous inequalities we did not see that J_{x_0} contains singularities of $w(z)$ but due to continuity of $\varphi(z)$ in the closed disk, these J_{x_0} do not affect the estimates obtained.

Replacing x by y in the previous arguments we similarly find

$$\int_{l_y(\gamma(t))} |\varphi(z)|ds \leq \frac{2}{\sqrt{2}\sin\alpha} \int_{-r}^{r} \int_{J_{y_0}} \left| \left(\varphi(z) \sin\arg\frac{\partial}{\partial y}w(z) \right)'_y \right| dy\,dx$$

$$+ \frac{\sqrt{2}}{2} \int_{-r}^{r} \max_{J_y} |\varphi(z)|dx := \Phi_y\left(r, \varphi(z)\right).$$

Taking into account that $l_x(\gamma(t))\bigcup l_y(\gamma(t)) = w^{-1}(\gamma(t))$ finally we obtain

$$\int_{w^{-1}(\Gamma)} |\varphi(z)|ds \leq \Phi_x(r, \varphi(z)) + \Phi_y(r, \varphi(z)). \tag{5.3.5}$$

5.3.4. An evaluation method for the number of zeros of complex functions. Now suppose that we have in the disk $D(r)$ a complex function $W(z) = U(x, y) + iV(x, y)$ satisfying the following conditions:

(i) the number $n(r, 0, W)$ of zeros of $W(z)$ in $\overline{D}(r)$ is finite;

(ii) w-images of smooth curves (as well as the pre-images of smooth curves) are piecewise smooth curves, and their smoothness can be broken only in a finite number of points $W(z^*)$, $z^* \in \overline{D}(r)$;

(iii) $V(x, y) = \operatorname{Im} w(z)$, where $w(z) \in M_\alpha(D(r))$, $\alpha > 0$ (see Section 1.4).

Our requirement for $U(x, y)$ is to guarantee the existence of the integrals below.

We note that any zero of the function $W(z)$ is an intersection point of some level sets of $U(x, y)$ and $V(x, y)$. However, the level sets of the function $V(x, y)$ actually coincide with the sets $w^{-1}(\gamma(0))$.

Let N be the number of connected components l_τ of the set

$$w^{-1}(\gamma(0)) \bigcap D(r) \backslash \{z_\tau^*\}. \quad \tau = 1, 2, \ldots, N.$$

We shall evaluate the number $n_\tau(r, 0, W)$ of zeros of $W(z)$ on l_τ. Let $\Psi(x, y)$ be a continuously differentiable function of two variables, such that $\Psi(x, y) > 0$, $\Psi_x'(x, y) > 0$ and $\Psi_y'(x, y) > 0$ for $(x, y) \in \overline{D}(r)$. Denoting the direction of the tangent to the arc l_τ by t and using the estimate (5.3.2) we obtain

$$n_\tau(r, 0, W) \leq \frac{1}{\pi} \int_{l_\tau} |\varphi(z)| ds + 1,$$

where

$$\varphi(z) = \frac{\partial}{\partial t} \arg \frac{\partial}{\partial t} \left[U(x, y) + i\Psi(x, y) U(x, y) \right].$$

But $\bigcup_\tau l_\tau = w^{-1}(\gamma(0))$ and $n(r, 0, W) = \sum_\tau n_\tau(r, 0, W)$, therefore, we get the desired estimate:

$$n(r, 0, W) \leq \frac{\Phi_x(r, \varphi(z)) + \Phi_y(r, \varphi(z))}{\pi} + N. \tag{5.3.6}$$

Thus, it is possible to find out geometric estimates for the number of zeros of very wide classes of functions $W(z)$. We used very few properties of the function $W(z)$, therefore we could use only the most general geometric arguments for its research.

However, it would be natural to presuppose that in the future certain equations of the form (5.3.1) will take a special place in general approaches to some physical problems, due to their closeness to the practice or to their mathematical significance. This evidently will lead to the consideration of other non-geometric properties of such systems in the investigations of zeros of the arising functions $W(z)$.

5.4. On a-points of some non-holomorphic function

For a given function $w(z) = u(x,y) + iv(x,y)$ holomorphic in a domain D and continuously differentiable non-constant function $\varphi(x,y)$ in D, we consider now the function $F(z) = u(x,y) + i(v(x,y) - \varphi(x,y))$, which obviously is non-holomorphic in D.

It turns out that it is possible to compare the averaged number of a-points of the functions $w(z)$ and $F(z)$ under some restrictions on φ. Namely, the following theorem is true.

Theorem 1. *Let $w(z) = u(x,y) + iv(x,y)$ be a function holomorphic in a domain D and let $\varphi(x,y)$ be a real, continuously differentiable function in D, such that $|\varphi'_x(x,y)| \le K$ and $\varphi'_y(x,y) \le K$, $(x,y) \in D$, for a constant $K < \infty$.*

If the function $F(z) = u(x,y) + i(v(x,y) - \varphi(x,y))$ realizes an interior mapping of D (see Section 1.5.1), then

$$\int_{-\infty}^{+\infty} [n(D, u = t, v - \varphi = m) - n(D, u = t, v = m)]dm$$
$$\le KL(D, \gamma^*(t), w), \tag{5.4.1}$$

where $n(D, u = t, v = m)$ is the number of $t + im$-points of the function $u + iv$ in D and $\gamma^*(t) = \{w : \operatorname{Re} w = t\}$. Besides,

$$\int_{-\infty}^{+\infty}\int_{-\infty}^{+\infty} [n(D, u = t, v - \varphi = m) - n(D, u = t, v = m)]\, dm dt$$
$$\le KS_1(D, w) \le K\, [S(D)S_2(D, w)]^{1/2}, \tag{5.4.2}$$

where S, S_1 and S_2 are those defined in Section 1.1.

Proof. Denote the $\gamma^*(t)$-lines of the function $w(z)$ in D by $l(u = t)$ and suppose that for a given t there are no multiple points of $w(z)$ on $l(u = t)$. Further, denote the connected components of the set of curves $l(u = t)$ by $l_i(u = t)$. Mapping $l_i(u = t)$ by the function $F(z)$ we obtain a set $l_i^F(u = t)$. On the other hand, the w-image of the arc $l_i^w(u = t)$ is an interval J_i on the straight line $\gamma^*(t)$. If τ is the direction of the tangent to a curve $l_i(u = t)$, then $\partial/\partial\tau u(x,y) = 0$ on $l_i(u = t)$. Consequently

$$|l_i^w(u = t)| := \int_{l_i(u=t)} |w'(z)|ds = \int_{l_i(u=t)} \left|\frac{\partial}{\partial\tau}v(x,y)\right| ds = |J_i| = \int J_i dm.$$
$$\tag{5.4.3}$$

Denoting the number of $t + im$-points of the function $u + iv$ on the set X by $n(X, u = t, v = m)$, we observe that $n(l_i(u = t), u = t, v = m)$ equals to 1 for $z \in l_i(u = t)$ and to 0 for $z \notin l_i(u = t)$. Therefore

$$|l_i^w(u = t)| = \int_{-\infty}^{+\infty} n(l_i(u = t), u = t, v = m)dm. \qquad (5.4.4)$$

Note that we did not consider those t for which there are multiple points on $l_i(u = t)$, as they do not change the integral. Further,

$$|l_i^w(u = t)| = \int_{l_i(u=t)} \left| \frac{\partial}{\partial \tau} v(x, y) - \frac{\partial}{\partial \tau} \varphi(x, y) \right| ds. \qquad (5.4.5)$$

Arguing as in proof of Theorem 1 of Section 5.4.5, we find

$$|l_i^F(u = t)| = \int_{-\infty}^{+\infty} n(l_i(u = t), u = t, v - \varphi = m)dm. \qquad (5.4.6)$$

From (5.4.3) to (5.4.6) it follows that

$$\int_{-\infty}^{+\infty} n(l_i(u = t), u = t, v - \varphi = m)dm - \int_{-\infty}^{+\infty} n(l_i(u = t), u = t, v = m)dm$$

$$\leq \int_{l_i(u=t)} \left| \frac{\partial}{\partial \tau} \varphi(x, y) \right| ds \leq K|l_i(u = t)|.$$

Summing up this inequalities over all i, we come to (5.4.1). Integrating (5.4.1) and using the estimate (5.4.14) of Section 1.1, we obtain (5.4.2).

REFERENCES

Ahlfors, L.

1. Untersuchungen zur Theorie der konformen Abbildungen und ganze Funktionen, Acta Soc. Sci. Fenn., v. 1, n. 9, 1930, pp. 1–40.

2. Beiträge zur Theorie meromorphen Funktionen, C.R.7. Congress des mathematicient scand, Oslo, 1929, pp. 84–88.

3. Zur Theorie der Überlagerungsflächen, Acta mathematica, v. 65, 1935, pp. 157–194.

Alt, H.W. and Caffarelli, L.A.

1. Existence and regularity for a minimum problem with free boundary, J. Reine Angew. Math., v. 325, 1981, pp. 434–448.

Andreian-Cazacu, C.

1. Uber die normal ausschöpfbaren Riemannschen Flächen, Math. Nachr., v. 15, 1956, Heft 2, pp. 77–86.

Arakelian, N.U.

1. Entire functions of finite order with infinite set of deficient values, Dokl. Acad. Nauk SSSR, v. 170, n. 2, 1966, pp. 999–1002 (in Russian).

Astala, K., Fernandez, J. and Rohde, S.

1. Quasilines and Hayman-Wu theorem, Indiana University Math. Journal, v. 42, n. 4, 1993, pp. 1078–1100.

Barsegian, G.A.

1. Deficient values and the structure of covering surfaces, Izvestia Acad. Nauk Armenii, v. 12, 1977, pp. 46–53 (in Russian).

2. Distribution of sums of a-points of meromorphic functions, Dokl. Acad. Nauk SSSR, v. 234, n. 4, 1977, pp. 761–763 (in Russian, translated in Soviet Math. Dokl.).

3. A Geometric approach to the problem of ramifications of Riemann surfaces, Dokl. Acad. Nauk SSSR, v. 237, n. 4, 1977, pp. 761–763 (in Russian, translated in Soviet Math. Dokl.).

4. To the distribution of distortions under mapping by meromorphic functions, Dokl. Acad. Nauk SSSR, v. 237, n. 5, 1977, pp. 1009–1011. (in Russian, translated in Soviet Math. Dokl.).

5. On geometric structure of image of disks under mappings by meromorphic functions, Math. Sbornic, v. 106(148), n. 1, 1978, pp. 35–43 (in Russian, translated in Math. USSR Sbornic).

6. New results in the theory of meromorphic functions, Dokl. Acad. Nauk SSSR. v. 238, n. 4, 1978, pp. 777–780 (in Russian, translated in Soviet Math. Dokl.).

7. On distribution of zeros of imaginary parts of meromorphic functions, Math. Zametki, v. 24, n. 2. 1978, pp. 183–194 (in Russian, translated in Soviet Math. Zam.).

8. Distribution of sums of a-points and Riemann surfaces of the class F_g, Math. Zametki, v. 25, n. 1, 1979, pp. 51–59 (in Russian, translated in Soviet Math. Zam.).

9. On the geometry of meromorphic functions, Mathem. sbornic, v. 114(156), n. 2. 1981, pp. 179–226 (in Russian, translated in Math. USSR Sbornic).

10. Exceptional value, associated with logarithmic derivatives of meromorphic functions, Izvestia Acad. Nauk Armenii, Ser. Mathematica, 16, v. 5, 1981, pp. 408–423 (in Russian, translated in Soviet Journal Cont. Math. Anal., Allerton Press).

11. About mutual disposition asymptotic paths and a-points of meromorphic functions, Izvestia Acad. Nauk Armenii, Ser. Mathematica, v. 18, n. 2, 1983. pp. 124–133 (in Russian, translated in Soviet Journal Cont. Math. Anal., Allerton Press).

12. Proximity property of a-points of meromorphic functions. Mathem. Sbornic, v. 120(162), n. 1, 1983. pp. 42–63 (in Russian, translated in Math. USSR Sbornic).

13. Proximity property of a-points of meromorphic functions and the structure of one-sheeted domains of Riemann surfaces, Izvestia Acad. Nauk Armenii, v. 20, n. 5, 1985. pp. 375–400; v. 20,

n. 6, 1985, pp. 407–427 (in Russian, translated in Soviet Journal Cont. Math. Anal., Allerton Press).

14. Estimates of derivatives of meromorphic functions on sets of a-points, Journal of London Math. Soc., v. 34, n. 2, 1986, pp. 543–400.

15. Tangent variation principle in complex analysis, Izvestia Acad. Nauk Armenii, v. 27, n. 3, 1992, pp. 37–60 (in Russian, translated in Journal Cont. Math. Anal., Allerton Press).

16. Principle of partitioning of meromorphic functions, Math. Montisnigri, v. 5, 1995, pp. 18–26.

17. Principle of closeness of sufficiently large sets of a-points of meromorphic functions in the unit disk, Hokkaido Math. Journal, v. 26, n. 2, 1997, pp. 451–456.

Barsegian, G.A. and Sukiasian, G.A.

1. Proximity Property of a-points of meromorphic functions with simple zero and poles, alternating on some systems of rays, ARM-NIINTI, n. 16, 1989, pp. 1–46.

Barsegian, G.A. and Yang, C.C.

1. On the cross-road of value distribution, Γ-lines, free boundary theories and applied mathematics. Preprint.

Beardon, A.F. and Yang, C.C.

1. Open problems for the workshop on Value Distribution Theory and its applications, Hong Kong, July 1996, Izvestia Nat. Akad. Nauk Armenii. Matematika, v. 32, n. 4, 1997, pp. 8–38 (in Russian, translated in Journal Cont. Math. Anal., v. 32, n. 4, 1997, pp. 9–36, Allerton Press).

Beurling, A.

1. On the free boundary problem for the Laplace equation, Sem. on Analytic Functions, Institute for Advanced Study Princeton, 1957, pp. 248–263.

References

Bieberbach, L.

1. *Über* eine Verteilung des Picardschen Satzes bei ganzen Funktionen endlicher Ordnung. Mathem. Zeitschrift, v. 3, 1919, pp. 175–190.

Bishop, C., Carleson. L., Garnett. J. and Jons. P.

1. Harmonic measures supported on curves. Pacif. Journal Math., v. 138, 1989, pp. 233–236.

Bishop, C. and Jons. P.

1. Harmonic measure and arclenght. Ann. Math., v. 132, 1990, pp. 511–547.

Borel, E.

1. Sur les zeros des fonctions entieres. Acta math., Stochh., v. 20, 1897, pp. 357–396.

Caffarelli, L.A. and Vazquez, J.L.

1. A free boundary problem for the heat equation arising in flame propagation. Trans. Amer. Math. Soc., v. 347, 1995, pp. 749–786.

Courant, R.

1. Dirichlet's principle. conformal mappings. and minimal surfaces. New York. 1950.

Courant, R. and Hurwitz. A.

1. Theory of functions. Moscow. Nauka. 1968 (in Russian).

Dinghas, A.

1. Vorlesungen *über* Functionentheorie, Die grundlegenden der Math. Wiss. in Einzehldarstellungen, B. 110, Springer-Verlag, Berlin–Göttingen–Heidelberg, 1961.

2. Wertverteilung meromorpher functionen in ein- und mehrfach zusammenhangenden Gebieten, Lect. Notes in Math., B. 783, Springer-Verlag, Berlin–Heidelberg–New York, 1980.

Edrei, A.

1. Meromorphic functions with three radially distributed values, Trans. Amer. Math. Soc., v. 78, n. 2, 1955, pp. 276–293.

Edrei, A. and Fuchs, W.

1. On the growth of meromorphic functions with several deficient values, Trans. Amer. Math. Soc., v. 93, n. 2, 1959, pp. 292–328.

2. Bounds for the number of deficient values of certain classes of meromorphic functions, Proc. London Math. Soc., v. 12, 1962, pp. 315–344.

Edrei, A., Fuchs, W. and Hellerstein, S.

1. Radial distribution and deficiencies of the values of a meromorphic functions, Pacif. J. Math., v. 2, 1961, pp. 135–151.

Evgrafov, M.A.

1. Analytic functions. Moscow, Nauka, 1968 (in Russian).

Federer, H.

1. Geometric Measure Theory, Springer-Verlag, New-York, 1969.

Fernandez, J.

1. Domains with strong barrier, Rev. Mat. Iberoamericana, v. 5, n. 1–2, 1989, pp. 47–65.

Fernandez, J. and Hamilton, D.

1. Length of curves under conformal mappings, Comm. Math. Helf., v. 62, 1987, pp. 122–134.

Fernandez, J. and Zinsmeister, M.

1. Ensembles de niveau des representations conformes, C. R. Acad. Sci. Paris, t. 305, Serie 1, 1987, pp. 449–452.

Fernandez, J., Heinonen, J. and Martio, O.

1. Quasilines and conformal mappings, Journal d' Analyse Math., v. 52, 1989, pp. 117–132.

Fuchs, W.H.J.

1. A theorem on the Nevanlinna deficienciences of meromorphic functions of finite order, Ann. Math., v. 68, n. 2, 1958, pp. 203–209.

Garnett, J., Gehring, F. and Jones. P.

1. Conformally invariant length sums, Indiana Univ. Math. Journal, v. 32, 1983, pp. 809–829.

Gelfond, A.O.

1. Über die harmonische Funktionen, Trudy Phys. Math. Instituta AN SSSR, n. 5, 1934, pp. 149–158.

Goldberg, A.A.

1. On meromorphic functions with divided zero and poles, Informations of higher educational institutions, Mathematicians, v. 4, 1960, pp. 67–72 (in Russian).

2. Meromorphic functions, Itogi Nauki i Texniki, ser. math., v. 10, 1973 (in Russian).

3. About integrals of the differential equations of the first order, Ukrainskiy Math. Journal, v. 8, 1956, pp. 254–261 (in Russian).

Goldberg, A.A. and Ostrovskii, I.V.

1. Distribution of values of meromorphic functions. Moscow, Nauka, 1970 (in Russian).

Hardt, R.

1. An introduction to geometric measure theory, Lecture Notes, Melbourne University, 1979.

Hayman, W.

1. Multivalent functions. Cambridge, 1958.

2. Meromorphic functions. Oxford, 1964.

3. An inequality for real positive functions, Proc. Cambridge Philos. Soc., v. 48, 1952, pp. 93–105.

4. Research problem in functions theory, London Math. Soc., Lect. Note ser., v. 12, 1974, pp. 143–154.

Hayman, W. and Wu, J.M.

1. Level sets of univalent functions, Comment. Math. Helv., v. 56, n. 3, 1981, pp. 366–403.

Hellerstein, S.

1. The distribution of values of a meromorphic functions and theorem of I.S. Wilf, Duke Math. Journal, v. 32, n. 4, 1965, pp. 749–764.

Hellerstein, S. and Korevaar, J.

1. The real values of an entire functions, Bull. Amer. Math. Soc., v. 70, 1964, pp. 608–610.

Kakutani, S.

1. On the exceptional values of meromorphic functions, Proc. Phys.-Math. Soc. Japan, v. 17. 1935. pp. 174–176.

Lelong-Ferrand, J.

1. Representation conforme et transformations a integral de Dirichlet bornee. Paris, 1955.

Lions, J.L.

1. Some methods of resolutions of free surface problems, Lect. Notes Physics, v. 59, 1976. pp. 1–31.

Markushevich, A.I.

1. Theory of functions of complex variable. Moscow, Nauka, 1967 (in Russian).

Miles, J.

1. Bounds on the ratio $sup\ n(r,a)/A(r)$ for meromorphic functions, Trans. Amer. Math. Soc., v. 162. 1971, pp. 383–393.

Miles, J. and Townsend, D.

1. On the imaginary values of meromorphic functions, Lect. Note Math. 599, 1977. pp. 93–95.

2. Imaginary values of meromorphic functions, Indiana Univ. Math. J., v. 27. n. 3. 1978. pp. 491–503.

Nevanlinna, R.

1. Eindentige analytische Funktionen. Berlin, Springer, 1936.

2. Zur Werteverteilung eindeutiger analytischer Funktionen, Abhandlungen aus dem Math. Seminar der I Hamburgschen Universität, v. 8, n. 4, 1930, pp. 351–400.

3. Uniformisierung. Springer-Verlag, Berlin, 1953.

4. Über die Eigenschaften meromorphen Funktionen, Acta math., v. 46, 1925, pp. 1–99.

Ostrovski, I.V.

1. The evaluation of the deficiency of meromorphic function at which two values are allocated inside an angle, Izvestia Vusov. Mathematica, n. 2, 1960, pp. 138–148 (in Russian).

2. On a disposition of zeros of entire functions, close to the real axes, Izvestia Acad. Nauk SSSR, Mathematica, v. 25, n. 2, 1961, pp. 277–238 (in Russian).

Öyma, K.

1. Harmonic measure and conformal length, Proc. Amer. Math., Soc., v. 115, 1992, pp. 687–689.

Petrenko, V.P.

1. Growth of meromorphic functions. Visha Shkola, 1978 (in Russian).

Selberg, H.L.

1. Über eine Satz von Collingwood, Arch. Math. Naturv., v. 48, n. 9, 1944, pp. 119– 226.

2. Eine Ungleichung der Potentialtheorie und ihre Anwendung in der Theorie der meromorphen Funktionen, Comment. Math. Helv., 18, 1946, pp. 309–326.

Shimizu, T.

1. On the theory of meromorphic functions, Japanese J. Math., v. 6, 1929, pp. 119–171.

Stoilov, S.

1. Theory of functions of complex variable, v. 2, In. Lit., Moscow, 1962, (in Russian).

2. Lecons sur les principles topologigues de la theorie des functions analitigues. Gauthier-Villars, Paris, 1956.

Suvorov, G.D.

1. Generalized length–area principle in the theory of maps, Kiev, Nauk. Dumka, 1985 (in Russian).

Tumura, Y.

1. Sur une extension d'un theoreme de O. Teichmüller, Proc. J. Acad., Tokyo, v. 19, 1943, pp. 55–59.

Tsuji, M.

1. Potential theory in modern function theory. Tokyo, 1959.

Väisälä, J.

1. Bounded turning and quasiconformal mappings, Monatschefte Math., v. 11, 1991, pp. 233–244.

Wilf, H.S.

1. The argument of an entire functions, Bull. Amer. Math. Soc., 67, 1961, pp. 488–489.

Zeldovich, Y.B. and Frank-Kamenetskii, D.A.

1. The theory of thermal propagation of flames, Zh. Fizic.-Chim (comm. Phys-Chemistry), v. 12, 1938, pp. 100–105 (in Russian; english translation in "Collected works by Y.B. Zeldovich", vol. 1, Princeton Univ. Press. 1992).

Zeldovich, Y.B., Barenblatt, G.I.. Libovich, V.B. and Makhviladze, G.M.

1. The mathematical theory of combursion and explosions, Consultants Bureau, 1984.

Index

Milton Keynes UK
Ingram Content Group UK Ltd.
UKHW020031071024
449327UK00032B/3015